AF791739

Micro-nanoelectronics Devices

Series Editor
Robert Baptist

Micro-nanoelectronics Devices

Modeling of Diffusion and Operation Processes

Christian Gontrand

iSTE PRESS

ELSEVIER

First published 2018 in Great Britain and the United States by ISTE Press Ltd and Elsevier Ltd

Apart from any fair dealing for the purposes of research or private study, or criticism or review, as permitted under the Copyright, Designs and Patents Act 1988, this publication may only be reproduced, stored or transmitted, in any form or by any means, with the prior permission in writing of the publishers, or in the case of reprographic reproduction in accordance with the terms and licenses issued by the CLA. Enquiries concerning reproduction outside these terms should be sent to the publishers at the undermentioned address:

ISTE Press Ltd
27-37 St George's Road
London SW19 4EU
UK

www.iste.co.uk

Elsevier Ltd
The Boulevard, Langford Lane
Kidlington, Oxford, OX5 1GB
UK

www.elsevier.com

Notices
Knowledge and best practice in this field are constantly changing. As new research and experience broaden our understanding, changes in research methods, professional practices, or medical treatment may become necessary.

Practitioners and researchers must always rely on their own experience and knowledge in evaluating and using any information, methods, compounds, or experiments described herein. In using such information or methods they should be mindful of their own safety and the safety of others, including parties for whom they have a professional responsibility.

To the fullest extent of the law, neither the Publisher nor the authors, contributors, or editors, assume any liability for any injury and/or damage to persons or property as a matter of products liability, negligence or otherwise, or from any use or operation of any methods, products, instructions, or ideas contained in the material herein.

MATLAB® is a trademark of The MathWorks, Inc. and is used with permission. The MathWorks does not warrant the accuracy of the text or exercises in this book. This book's use or discussion of MATLAB® software or related products does not constitute endorsement or sponsorship by The MathWorks of a particular pedagogical approach or particular use of the MATLAB® software.

For information on all our publications visit our website at http://store.elsevier.com/

© ISTE Press Ltd 2018
The rights of Christian Gontrand to be identified as the author of this work have been asserted by him in accordance with the Copyright, Designs and Patents Act 1988.

British Library Cataloguing-in-Publication Data
A CIP record for this book is available from the British Library
Library of Congress Cataloging in Publication Data
A catalog record for this book is available from the Library of Congress
ISBN 978-1-78548-282-3

Contents

Preface

Microelectronics extends from materials to systems architecture, encompassing the "process", the "device" and the circuit. For many years now, nanoelectronics have been the subject of discussion in industries as well as within concerned public entities. This is reality, even though the latest fashion, and, more pragmatically, the ongoing search for contracts, means that this prefix is ubiquitous. Each specialization has its own definition for this term. For the time being, let us say that nanoelectronics concerns electronic systems whose devices contain at least two "strongly" sub-micronic dimensions, typically less than 100 nm.

In the 1980s, the modeling of diffusion (process) and the electric behavior of components (device) became a reliable practice in microelectronics.

Due to the reduction in the size of devices, unidimensional numerical simulations are no longer relevant. This may not always be true for analytical solutions of equations.

There are many publications or books in the field of microelectronics; in this book, we concentrate mainly on the "process" and the "device" when it comes to modeling. The semiconductor material of reference will be silicon, which is by far the most widely used material. Thus, our aim is not to provide an exhaustive study, but to focus on the heart of the modeling of integrated devices:

– we will elaborate here on the "upstream" – that is, we will introduce the equations that underpin models and, *in fine*, related simulations;

– regarding "downstream", we will try to propose some modifications that must be carried out, as regards modeling, to enter the field of nanoelectronics wholeheartedly.

Although this book is not dedicated to micro-devices and nanoelectronics as a whole, there are several books on this subject that can be referred to, ranging from real technological advances to those that are ironically called “grunge devices”. We will offer a succinct review of the family of current micro(nano)electronic devices.

From a practical point of view, we will propose some simple computer programs for an initiation in numerical analysis; being highly simplified, it can serve as the key to simulators that make it possible to carry out “numerical experiments” and extend these algorithms in new directions, especially by developing new models, namely in practice, introducing new equations or even new relations between equations.

We hope that the ideas introduced here may help the reader to get a better grasp of this domain and gain a certain perspective when they read other related research articles. Certain questions remain open; higher-level mathematicians will explore these integro-differential physics equations, looking for possible bridges connecting them.

This book addresses researchers and students at the master’s and doctoral levels in a spirit of “teaching through research”, and the required level of mathematics is assimilated sophomore-level knowledge.

I wish to thank some of my colleagues, especially former or current doctoral candidates or master’s level students working on their thesis. This list is by no means exhaustive: Saïda Latreche, Christiane Dubois, Olivier Valorge, Fengyuan Sun, Samir Labiod, Mourad Bella, Antoine Bacha, Maxime Pirot, Damien Rabourdin, Jean Steenhouwer, Francis Calmon, Jacques Verdier, Pierre-Jean Viverge, Guy Chaussemy, Michel Perez, Daniel Barbier, Alain Poncet, André Laugier, Daniel Gasquet, Jean-Claude Vaissiere and Jean-Pierre Nougier.

Christian GONTRAND
January 2018

1

Introduction and Reminders

1.1. A brief history of microelectronics

1.1.1. *Electricity: Ampère, Coulomb, Faraday, Gauss, Henry, Kirchhoff, Maxwell, Ohm*

1826	Ohm's law (Georg Simon).
1837	S. Morse (New York): invention of the telegraph using binary signals (dots and dashes); also W. Thomson and C. Wheatstone (Playfair cipher: an encryption technique).
1876	Bell: invention of the telephone.
1877	Edison: development of the phonograph (disk: first ROM).
1865	J.C. Maxwell: formulation of electromagnetic theory.
	Hertz: production of electromagnetic waves in a laboratory.
1996	Marconi: development of the wireless telephone using Hertzian waves (km).

1.1.2. *Vacuum tube*

1895	H.A. Lorentz: theory of electrons (Greek: amber); discrete charges.

1897	J.J. Thomson: experiment conducted to account for the existence of electrons.
1897	Braun: development of the cathode ray tube and the first electronic tube.
1904	Ambrose Fleming: invention of the diode (tube) detector.
1905	Albert Einstein, Lorentz, Henri Poincaré: special relativity and unifying electromagnetism.
1906	Pickard: development of the cat's-whisker detector using a silicon (Si) crystal, which was not very reliable because of the tips.
1906	Lee de Forest: invention of the Audion triode (diode + grid: forerunner of the transistor), the first controlled source.

1.1.3. *Early applications*

1911	The triode was reliable (cathode covered by an oxide coat + excess vacuum). Telephony, radio communications.
1917	The IRE (Institute of Radio Engineers) was set up.
1884	AIEE (American Institute of Electrical Engineers).
1906	AIEE + IRE: IEEE (Institute of Electrical and Electronics Engineers). Diodes and triodes.
1912	Armstrong: cascading a negative-feedback amplifier. Lee de Forest: development of the oscillator.
1917	Armstrong: development of the heterodyne receiver (see translation of frequency).
1918	Eccles–Jordan: invention of multivibrators.

	Cascading positive-reaction amplification + heterodyne: detection of weak signals.
1930	Armstrong: development of frequency modulation, before amplitude modulation (AM).
1930	B&W television.
1950	Color television.
1942	Radar (Radio Detection and Ranging).
	Hyper frequencies/Klystron Magnetron.

1.1.4. *Computers (transistors–trans-resistors + integrated circuits (IC))*

1633	Wilhelm Schickard: invention of mechanical calculator, with wheels having varying numbers of spokes.
1943	Blaise Pascal.
1931	Gottfried Wilhelm Leibniz.
1883	Babbage: invention of the "Analytical Machine" using perforated cards (see 1853; Jacquard) to record programs.
1930	Aiken: invention of the "Harvard Mark I", the automatic calculator with programmed sequences (about 17 m * 3 m).
1936	Alan Turing: general principles of automatic state machines.
1945	IBM (Industrial Business Machines): introduction of the 603 (commercialized), the 701 in 1952 and the 704 in 1954 (144 kb of memory).
1945	Von Neumann: theory of the "architecture of automatic calculators".
1946	Eckert, Mauchly (Pennsylvania): ENIAC (Electronic Numerical Integrator and Computer).

	Military: ballistics, Von Neumann: binary.
	40*2,300 tubes (room: 10 m * 13 m).
1947	IBM 604: 4,000 copies in 12 years.
1948	Start of the computer industry: advent of the transistor (trans(re)sistor).
1951	UNIVAC 1: first commercial computer.
1954	IBM 650: first-generation digital computer.
	Bush (Massachusetts Institute of Technology): creation of the differential analyzer, the first electromechanical analog computer.
	Operational amplifier: analog electronic.
1955	SABRE: first informatics network (created for American Airlines).
	W. Shockley quit Ma Bell to set up his own company in Palo Alto, California, the first of what would become Silicon Valley.
1956	Cray: founded the Control Data Corporation, then Cray Computers.
	Computers with semiconductors.
1957	John Backus: invention of the first "high-level" programming language FORTRAN.
1959	IBM 5090/7094: second generation.
	PDP 1: first interactive computer (from the Digital Equipment Corporation).
	PDP8, 1965: first mini-computer in the industry.
1964	IBM 360: hybrid ICs with discrete transistors on a (bulk).
	Burroughs Control DATA, UNIVAC.
1970	IBM 370: third generation.

1980	CI VLSI (Very-Large-Scale Integration): fourth generation. Tens of millions of operations/second. New architectures (vector, pipeline).
2010	Hexaflop Teles computer.

1.1.5. *Analysis and theory*

Analysis of circuits, synthesis technologies

BELL & MIT

1938	Bode, Nyquist: feedback amplifiers.
1940s	Shannon, Kotelnikov, Spataru: information theory (transmission of data).
1938	E.g.: invention of PCM (Pulse Code Modulation) by Reeves.
1937	Shannon (1937): Boolean algebra; analysis and design of communication circuits.
1936	Turing: introduced the concept of a universal machine.
1952	Wilkes: introduced the concept of microprogramming.
1940	Raggazini, Zadeh: sampled data systems (digital control and commands).

1.1.6. *Transistor*

End of 1947	Bardeen, Brattain, Shockley (Bell Telephone): designed the bipolar transistor.
1930 1945	Study of the electromagnetic properties of semiconductors and metals.

	Block, Davidov, Lark, Horovitz, Mott, Schottky, Slater, Summerfield, Vanvleck, Wigner, Wilson, van der Ziel.
1950	Teal (Bell Labs): research division of AT&T.
	Drawing ultra-pure germanium (Ge) monocrystals (Czochralski method).
1951	Commercial production of transistors.
	ATT issued patent permits for the manufacture of transistors: RCAn Raytheon, General Electric, Westinghouse, Western Electric (the manufacturing branch of ATT).
1954	Teal (Texas Instrument): production of silicon (Si) transistors.
1956	Bardeen, Brattain, Shockley: Nobel Prize.
1975	Esaki: bandgap engineering.

1.1.7. *Integrated circuits*

1958/9	Kilby (Texas Instruments): IRE (Institute of Radio Engineers) Congresses: semiconductor circuits (multivibrators, oscillators, Si or Ge capacitors).
	Noyce (Fairchild): future founder of Intel.
	Monolithic Si circuits; many devices: resistance, capacitors, d-isolation.
	PN junction for insulation, patented by Lehovec (Prague Electric Company).
	Moore: diffused zones created into silicon.
1958	Hoerni (Fairchild): diffused transistors (the base and the emitter diffused in the collector).
	Passivization of junctions using oxide layers.
	Lithography, etching.

Treatment by lot (several chips (dies) on the same wafer).

IC was commercialized by Texas and Fairchild.

1.1.8. *Field-effect transistor*

1951 Shockley: JFET (Junction Field-Effect Transistor).

Pb; unstable surface (traps for charge carriers).

1958 Tezner (France): manufacture of the first JFET, resulting from the planar procedure (replacing the mesas, plateaus and S_iO_2 passivization).

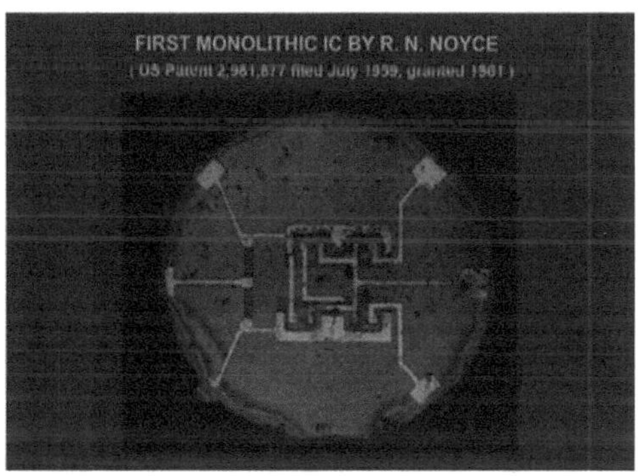

Figure 1.1. *1961: Technology Planar by Robert NOYCE @ Fairchild Inc. For a color version of this figure, see www.iste.co.uk/gontrand/nanoelectronics.zip*

1961 Robert N. Noyce (Fairchild, Inc.): invention of planar technology.

1960 Atalla, Kahng (Bell Labs): development of the first MOS (Metal Oxide Semiconductor), p-type MOS (PMOS).

1962 First CMOS inverter.

1962 Hofstein, Heiman (RCA): obtained the patent for the manufacture of IC MOS, the first commercial field-effect transistors.

1970 CMOS (Complementary MOS): MOS and PMOS.

1970	BiCMOS (Bipolar Compatible CMOS): e.g. the polycrystalline NPN emitter, produced at the same time as the polysilicon of the PMOS.
1951	Discrete transistors per chip.
1960	Small-scale integration (100 < devices < 1,000).
1966	Medium-scale integration (100 < devices < 1,000).
1969	Large-scale integration (1,000 < devices < 10,000).
1975	Very-large-scale integration – VLSI (10,000 < devices).
1986	Ultra-large-scale integration – ULSI (1,000,000 < devices).
2010	Towards 3D?

1.1.9. *Digital integrated circuits*

1961	Buie: transistor-coupled transistor logic (Pacific Semiconductor, ∈TRW).
	TTL (Transistor–Transistor Logic).
	E.g.: several emitters per transistor, stronger integration of devices.
1962	Motorola: ECL (Emitter Coupled Logic).
1972	Hart, Slob (Phillips, the Netherlands) and Berger, Wiedman (IBM, Germany): I^2L (Integrated Injection Logic); transistor with several collectors and very high density bipolar chips.
	CMOS: PMOS and NMOS (initial use: watches, e.g. in Japan and Switzerland).
1990	CMOS supremacy.
	MOS are widely used in RAM (Random Access Memory).

1.1.10. *Volatile memory (read and write)*

1970	Bipolar RAMs (storing about 1,000 bits (binary digit) of information).
1973	16,000 bits (MOS).
1978	64,000.
1982	288,000.
1988	$> 10^6$.
2000	ROM (Read Only Memory, 1967).
	PROM (Programmable ROM).
	EPROM (Erasable PROM).

1.1.11. *MOS*

1965	The company Intel was founded (Silicon Valley).
1968	The network ARPANET, the forerunner of the INTERNET was launched.
	Invention of the Windows–mouse environment (XEROS, which gave way to Microsoft).
1969	M.E. Hoof (Intel): microprocessor.
	UNIX/Operating System.
1971	Intel 4004: creation of the first microprocessor (*calibrated to 740 kHz*; 2,300 transistors).
1972	micro 8 bits.
1973	Programming language C was created, strongly linked to the UNIX OS.

1976	CRAY1: supercalculator launched (peak power of 100 MFlops).
1977	The first Apple computer.
1977	micro 16 bits. 32 bits, then 64 bits.
1981	Cochran and Boone (Texas): patented a "microcomputer" on a chip (however, the first to be available on the market was Intel 8048).

1.1.12. *CCD (MOS): charge-coupled device ("multi-gate" MOS)*

1970	Boyle, Smith (Bell Labs): memories and registers with 64,000-bit RAM.
1977	Application: cameras, image processing, telecoms.
1964	Wildar (Fairchild Semiconductor): first operational amplifier, the µA 709. Most analog circuits consist of bipolar transistors (see output speed and fan out) and also MOS from the late 1970s. E.g. – analog multiplicators (e.g. Gilbert). – Digital, analog and A/N. – PLL (Phase-Locked Loop) and its VCO (Voltage-Controlled Oscillator).

1.1.13. *Manufacturing technique*

1960	Epitaxy (Greek: *epi* meaning above and *taxis* meaning classification).
1967	Electron-beam masking.

1971	Ion implantation: I^2.
1981	IBM started marketing personal computers.
1983	Creation of the C++ programming language.
1984	Apple Macintosh launched the first successful marketing of a mouse–Windows environment (XEROS Technology).
1986	Microsoft launched the Operating System Windows 1.1.
1989	World Wide Web and HTML (Hyper Text Markup Language) launched at the Centre Européen de Recherche Nucléaire (CERN).
1994	Intel launched the Pentium, a microprocessor containing over 5 million transistors. Novel features: data buses enlarged to 64 bits for access to memory, and the processor can process two instructions by the clock cycle and two levels of memory cache in order to accelerate instruction processing at processor level.
1998	Intel's Pentium II and AMD's K6-2 were launched.
2001	Launch of the Intel Pentium III. The processor increased the frequency of PCs to 866 MHz.
2003	Launch of the Intel Pentium IV. With an initial frequency of 1 GHz, it would later increase to a frequency 6.4 GHz.
2003	The number of transistors on a PC chip increased up to billions. A PC microprocessor can deliver up to 6.4 GFlops.

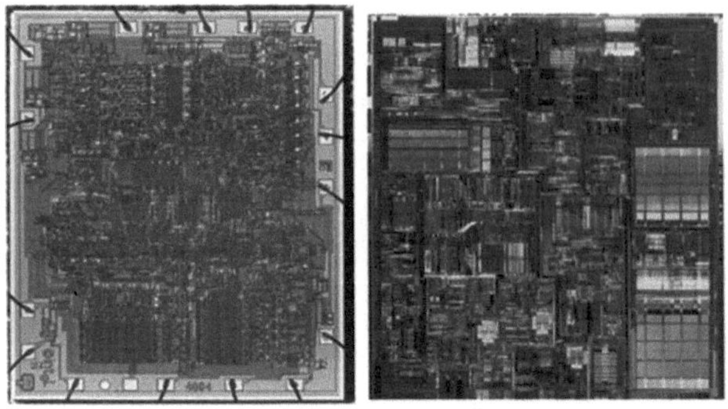

(a) (b)

Figure 1.2. *The first microprocessor: Intel 4004 (1971); Intel P4 (2000). (a) The first microprocessor: Intel 4004 (1971); 2,300 transistors; 0.63 Watt; NMOS: LG = 10 μm; Fmax= 740 kHz; 93,000 op/s. (b) Intel P4; 42 million transistors; 72 Watts; NMOS: LG = 0.18 μm; Fmax= 2 GHz;9.7 million op/s. The most powerful supercomputer that can reach operating speeds in the peta FLOPS range is IBM's DOE Blue Gene, set up at the Lawrence Livermore National Laboratory (USA). For a color version of this figure, see www.iste.co.uk/gontrand/nanoelectronics.zip*

1971 Ion implantation: I^2.

Interconnection lengths on each chip:

1961 25 μm.

1975 2 μm.

1990 1 μm.

1.1.14. *CAD: computer-aided design*

Manufacturing process simulation

SUPREM (Stanford) 1978: first 1D simulator

BICEPS ZD (finite difference discretization)

TITAN (CNET Grenoble 3D finite elements: process), ATHENA SILVACO

For the device:

PISCES/MEDICI

STORM (European project)

ATLAS SILVACO

SENTAURUS SYNOPSIS

Electric simulator circuits (Kirchhoff laws)

SPICE: Simulation Program for Integrated Circuits Emphasis (Vladimirescu).

ELDO: very well adapted to the CMOS and RF; electric simulator RF:

Layout/Schematic: CADENCE

ADS (Advanced Design System: Hewlett Packard).

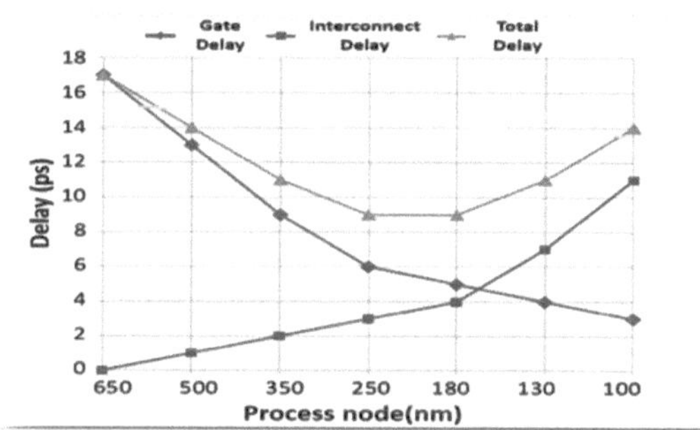

Figure 1.3. *Delay time as a function of the process node dimension. For a color version of this figure, see www.iste.co.uk/gontrand/nanoelectronics.zip*

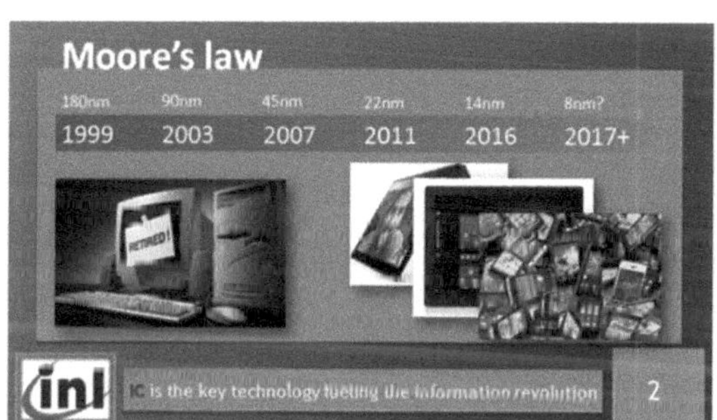

Figure 1.4. *Moore's law (1975): the number of transistors on the microprocessors (and no longer simple integrated circuits, as was the case with the first conjecture of 1965) on a silicon chip doubles every 2 years. For a color version of this figure, see www.iste.co.uk/gontrand/nanoelectronics.zip*

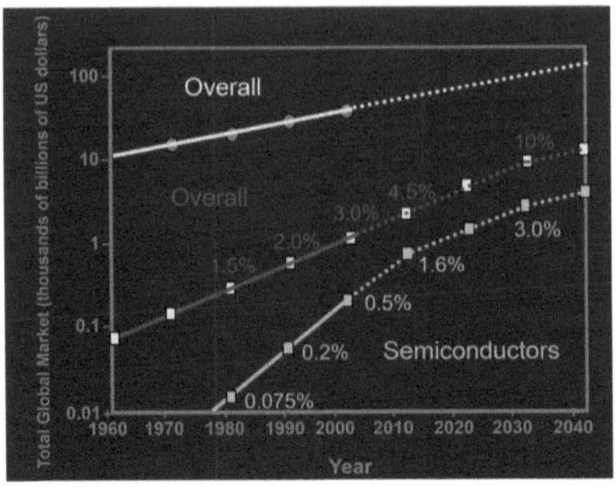

Figure 1.5. *Semiconductor industry. For a color version of this figure, see www.iste.co.uk/gontrand/nanoelectronics.zip*

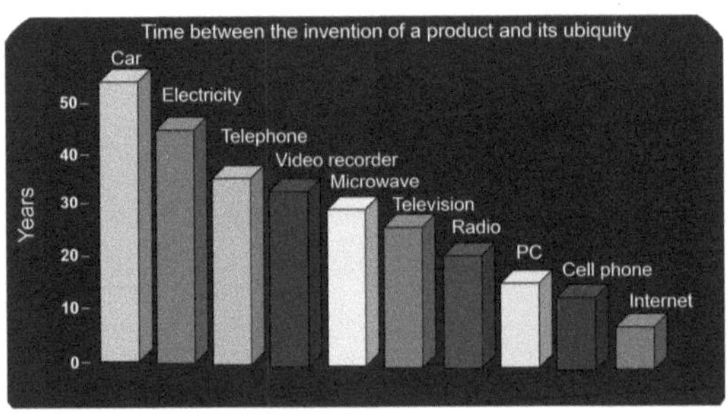

Figure 1.6. *Penetration into the market. For a color version of this figure, see www.iste.co.uk/gontrand/nanoelectronics.zip*

The position of Computer Aided Design (CAD)

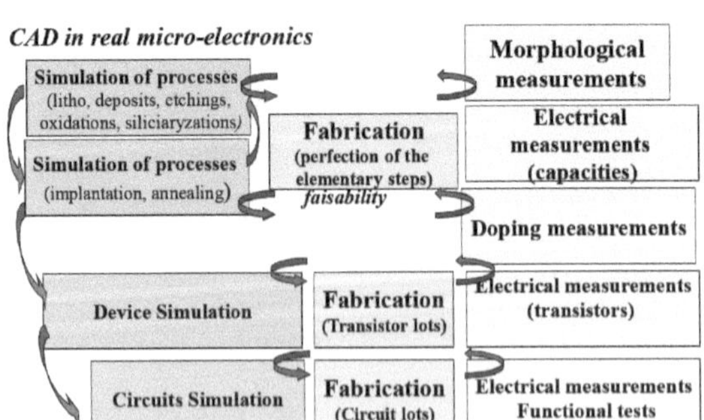

Figure 1.7. *CAD: Computer-Aided Design. For a color version of this figure, see www.iste.co.uk/gontrand/nanoelectronics.zip*

1.2. Technological computer-aided design

Commands	Parameters		
– Specification of models	Technological	Physical	Numerical
– Description of substrate	Thickness	Anisotropy coefficient	Mesh size
– Definition of masks	Dimension	Reaction coefficient	Time step
– Deposit	Duration	Diffusivity	Number of iterations
– Etching	Dose	…	Convergence criteria
– Ion implantation	Energy		Relaxation parameter
– Annealing	Gas pressure		
– Electrical contacts	Temperature		
– Saving the results	…		

Table 1.1. *Simulators for technological procedures*

TCAD in microelectronics:

– 100% predictive technological simulation does not exist!

- Physical mechanisms have become increasingly complex.

- Technology is evolving rapidly; the field of modeling finds it hard to keep up!

- Modeling calibrations always need experimental data.

- In spite of highly rigorous monitoring of technical parameters in the silicon industry, fluctuations are never non-negligible!

– *Commercial* simulators are used more in the industry than in research.

These tools are:

– powerful at interpolating between well known situations;

– uncertain when extrapolating and exploring new concepts.

1.3. Manufacturing [WAR 66]

Diffusion processes are definitely the most crucial steps in the manufacture of integrated circuits. They are underpinned by the diffusion equation, which is the same as the heat equation [CRA 75, FAI 86, PHI 91, MAT 91, ADD 66, BOU 03].

Physicists and technologists came together in the 1970s to work on bidimensional problems (e.g. lateral diffusions, diffusion accelerated by oxidation or due to the inevitable presence of other doping agents); indeed, they worked so closely that their roles were hard to distinguish.

We must also keep in mind that the physics of the occasional defects, metallic defects, for example, in the context of semiconductors are far from being taken into account in commercial simulators: miniaturization may aggravate these problems.

Since the 1980s, ion implantation has replaced initial diffusions: the mean depth of the implanted doping agent is well controlled and the subsequent rapid thermal annealing (RTA) makes it possible to adjust the post-implantation diffusions (drive-in) [BIE 75, FAI 84, LIN 63].

The diffusion equation – for the dopants, here – is one of the fundamental equations in physics and involves the dreaded Laplacian operator!

Principal steps in the manufacture of devices are:

a) ion implantation: I^2;

b) redistribution (drive-in);

c) thermal oxidation;

d) epitaxial growth;

e) etching;

f) deposition, e.g. oxide deposits.

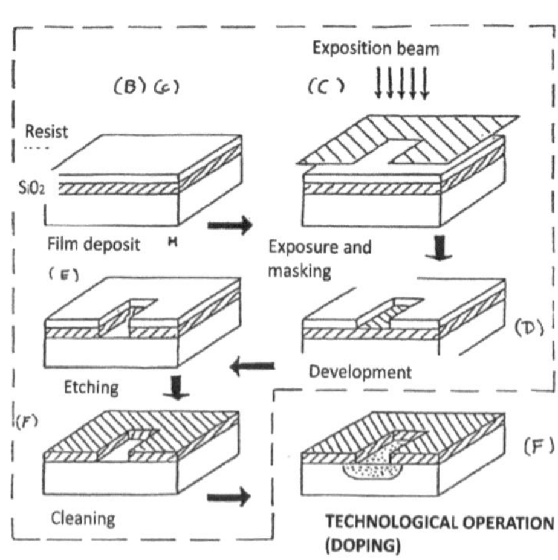

Figure 1.8. *The main steps in the process for creating a doped zone*

Until the 1980s, operation at high temperatures would last for tens of seconds or minutes. This then came down to several seconds using RTA, up to the point where it is often no more than a flash today! This reduces the redistribution or diffusion of doping agents. The distribution curve for the doping species will thus be modified due to thermal diffusion: under the effect of the temperature, doping impurities will move from sites of high concentrations to low concentrations. Within a few milliseconds at the start of the annealing, most of the doping agents go from interstitial (anti-sites) to substitutional (on-site), which then occupy the edges of the cubes (let us recall silicon: face-centered cubic), allowing them to be electrically active.

1.3.1. *Diffusion processes*

Diffusion processes depend on the nature of the source of doping agents. There are three main types of sources for the doping agents that are forced to penetrate into substrates. Historically, these sources are gaseous, liquid or solid.

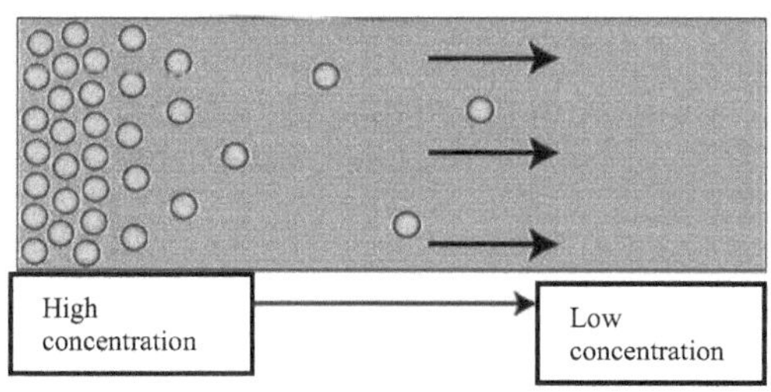

Figure 1.9. *Diffusion in an oven using gaseous sources. The doping gases are carried by a neutral carrier gas such as nitrogen*

Gaseous sources are gases such as arsine (AsH_3), phosphine (PH_3) or diborane (B_2H_6). Note that a neutral gas (nitrogen) always circulates in order to prevent any pollution by elements from the ambient atmosphere. This nitrogen must be very pure so as not to pollute the oven.

While these gases are, *a priori*, simple to use, they are very dangerous, as they are lethal to humans at a concentration of just a few ppm. Therefore, liquid sources, such as $POCl_3$ or BBr_3, are preferred. These sources are liquid at ambient temperature, but can easily be vaporized when introduced into diffusion ovens.

We can also use solid sources, which are glasses that contain doping agents such as boron nitride or glass doped with phosphorous. These sources are in the form of wafers, and, generally, these wafers and the wafers to be doped are introduced alternately into the oven. These wafers are activated by oxidation, which allows for the formation of an oxide. In the case of boron, BO is formed. These oxides or glasses evaporate and get deposited onto the wafers (to be doped). As they are being deposited, a fraction penetrates from the surface into the silicon. After the solid sources are removed from the oven, doping elements are diffused towards the interior of the substrate during a high-temperature thermal step. In practice, before the diffusion step, we eliminate the glass deposited by chemical etching, as the quantity of doping agent introduced onto the surface of the substrate during deposition is sufficient.

1.3.1.1. *Diffusion from gaseous sources (no longer widely used)*

In the case of gaseous sources, the surface concentration in the ambient medium is constant; this means that in the solid phase, the surface concentration, Cs, is also constant.

The limiting condition can thus be written as:

C(0, t)=Cs (whatever the time at x=0)

The second condition is an initial condition which assumes that the concentration of the substance to be diffused is initially null (or negligible) in the substrate. This condition is thus written as:

C(x, 0)=0 (whatever the value of x at t=0)

The last condition is more intuitive. We can well imagine that at an infinite distance, the concentration is null regardless of the time.

When the penetration depth of dopants is larger than the thickness of the initial surface film, the profile has a Gaussian form.

Upon integrating the differential equation for diffusion, with the diffusion (or diffusivity) coefficient D being constant, in 1D, [CRA 75],

$$\frac{\partial C}{\partial t} = D\,\frac{\partial^2 C}{\partial x^2} \qquad [1.1]$$

we have:

$$C(x,t) = \frac{Q}{\sqrt{\pi D t}} \exp\left(-\frac{x^2}{4Dt}\right) \qquad [1.2a]$$

where Q (the total quantity of deposited impurities) = $\int_0^{\infty} C(x)dx, \forall t$ [1.2b]

1.3.1.2. *Diffusion from solid sources or surface doses (no longer widely used)*

This is also the case when the dopant has been introduced onto the surface by ion implantation: I^2 (see below). The total quantity (or dose) is constant: C(x, 0) = 0 is the initial condition.

Let us recall that the total quantity of the dopant per unit of the surface (or the dose) is constant (if there is no exodiffusion). If we keep the concentration constant during the annealing, we have a profile of the type erfc (complementary):

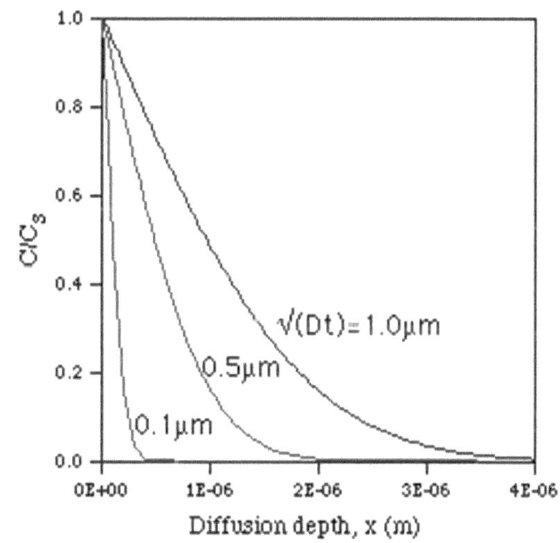

Figure 1.10. *Different profiles along time. For a color version of this figure, see www.iste.co.uk/gontrand/nanoelectronics.zip*

Doping is carried out in two distinct steps: ion implantation (since the 1970s, this process has replaced initial diffusions) and activation annealing.

The first step consists of accelerating doping impurities in order to incorporate them into the pure silicon (also called intrinsic silicon) at a specific dose (at/cm^2) and depth.

The second step makes it possible to repair damage caused by ion implantation (within a few milliseconds) and thus to restore the crystalline nature of the substrate (the most commonly used substrate is silicon).

Annealing also ensures that the dopant is activated by positioning it on the crystalline sites of the silicon (on the sites, also called substitutional); it also shows that an ideal matrix cannot be perfect.

1.3.2. *Diffusion modeling*

Diffusion modeling is now so well established that the heat equation or the diffusion equation may also be deduced from Brownian motion. The “random walk” is modeled based on random or stochastic movements – on the Internet, Google (see Page Rank) or in the street (a drunkard, for instance). We can also see that the hiccups in the present calculations make use of this, which may be of interest later on in this book.

Note that random microscopic phenomena may introduce deterministic macroscopic behavior.

In the case of silicon, which is an element in the fourth group in the periodic table, donors are elements from the fifth group (the most commonly used are phosphorous and arsenic), while acceptors are elements from the third group (the most commonly used are boron and indium).

On the atomic scale, a semiconductor has, at ambient temperature, a nearly full valence band and a nearly empty conduction band.

The valence band is the energy band in which electrons contribute to the local cohesion of the crystal by binding neighboring atoms (hence, the term “covalent bond”). Conduction electrons, or free electrons, occupy the zone called the conduction band. These electrons travel between atoms to generate electric current.

Conduction may also be carried out by “holes” or “vacancies” in atoms; that is, the lack of an electron “moving” into the valence layer, instead of an excess of atoms into the conduction layer.

Doping consists of introducing an excess of electrons by adding donor ions in the silicon and a deficit of electrons (and thus an excess of holes) by adding acceptor ions.

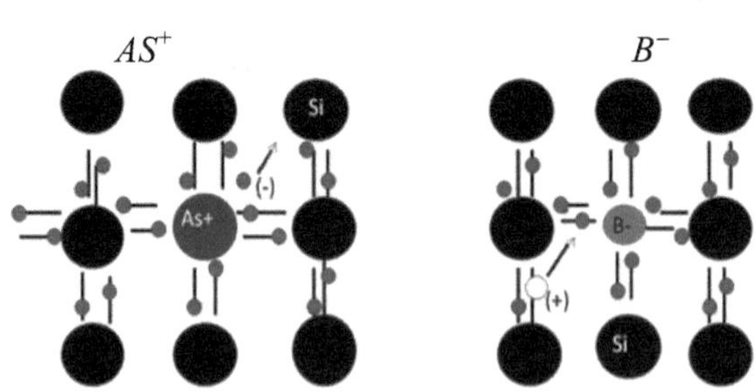

Figure 1.11. *Introducing: As dopant (red atom, on left) and B dopant (blue atom, on right) in a matrix of silicon (“black atoms”). For a color version of this figure, see www.iste.co.uk/gontrand/nanoelectronics.zip*

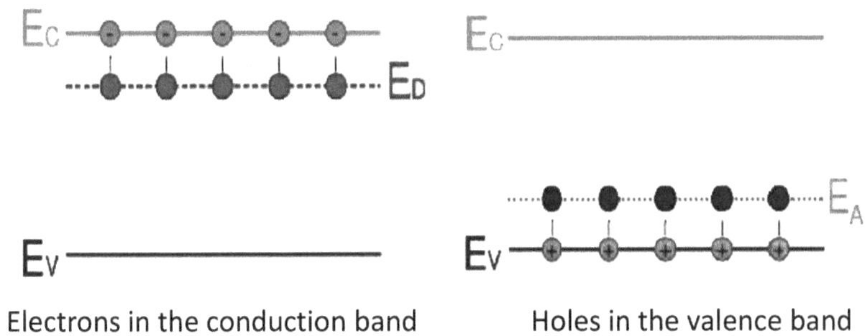

Figure 1.12. *Band diagram in doped silicon*. For a color version of this figure, see www.iste.co.uk/gontrand/nanoelectronics.zip*

P-N junctions realized through doping in silicon are essential to make transistors, the basic devices of microelectronic circuits.

1.4. PN junction (an overview)

DEFINITIONS.–

– *PN junction*: the semiconductor shifts from P-type to N-type within the same crystal.

– *Homojunction*: the P-type semiconductor is made up of the same material (Si, Ge, GaAs) as the N-type semiconductor. In opposite case, we talk of a heterojunction.

– *Metallurgic junction*: the plan/arrangement where the semiconductor changes the type.

– In the unidimensional model, the distribution of impurities is studied only along an Ox axis.

– *Impurity profile*: this is the difference between the density of acceptor atoms and the density of donors ($N_A - N_D$).

– To simplify things, we will assume that from the *"P" side*, the excess density of acceptor atoms is equal to $\mathbf{N_A}$ (cm^{-3}), and from the *"N" side*, the excess density of donor atoms is equal to $\mathbf{N_D}$ (cm^{-3}).

– *Step junction*: the shift from the "P" region to the "N" region takes place over an infinitely fine thickness.

– *Linearly graded junction*: the shift from the "P" region to the "N" region follows a linear law.

– In reality, the impurity profile is quite well represented by the erfc(x) function, which is called the *real junction.*

– *Junction:* different profiles of junction.

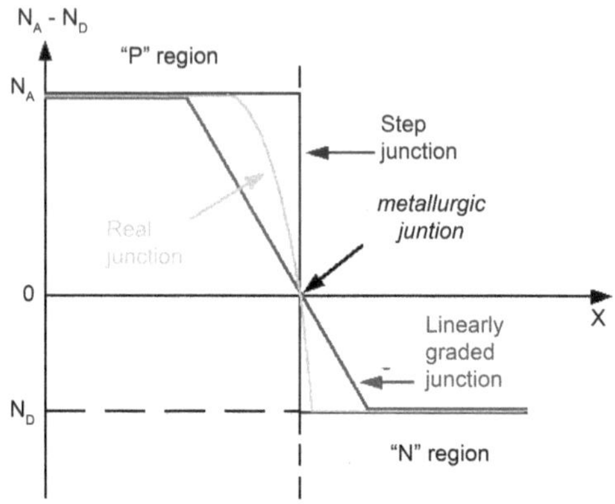

Figure 1.13. *Different junction profiles. For a color version of this figure, see www.iste.co.uk/gontrand/nanoelectronics.zip*

The *PN diode* is made up of a "P" semiconductor region, a PN junction and an "N" semiconductor region.

In practice, modern PN diodes are strongly asymmetrical. Doping in one zone is much greater than that in the other (to enhance the injection of carriers).

If the "P" region has more dopants than the "N" region, we talk of a P^+N *diode.*

If the "N" region has more dopants than the "P" region, we talk of a PN^+ *diode.*

With respect to a vacuum diode (see Fleming's diode), we refer to the "P" region of the diode as the *anode* and the "N" region as the *cathode.*

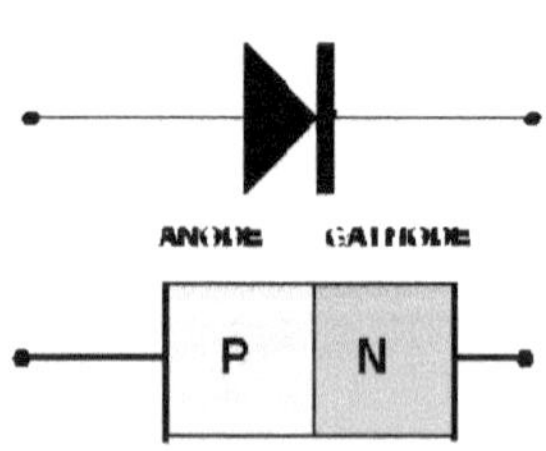

Figure 1.14. *The diode – (reflected) K in the scheme: cf. Kathod mnemonic – and we refer to the "N" region as the cathode*

The PN diode presents the distinctive feature of allowing the current to pass in only one direction.

1.4.1. *The non-polarized PN junction*

– Let there be a step PN junction (constant P doping = N_A, constant N doping = N_D) at thermodynamic equilibrium (ideal junction).

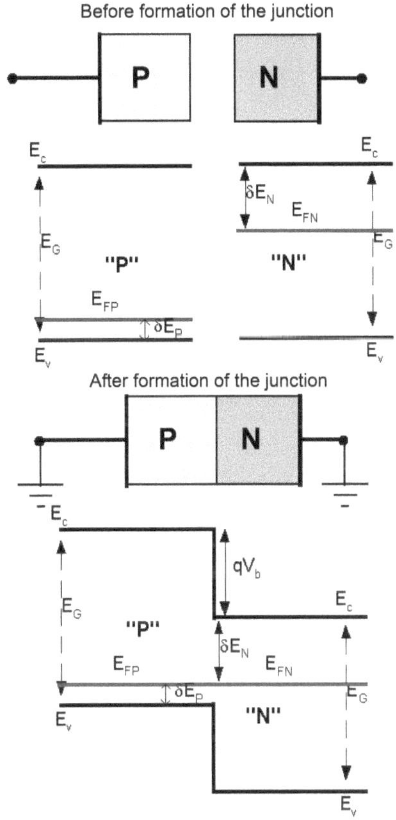

Figure 1.15. *Band diagram of a non-polarized PN junction. For a color version of this figure, see www.iste.co.uk/gontrand/nanoelectronics.zip*

– Let us imagine that the "P" semiconductor is separated from the "N" semiconductor.

In the "P" semiconductor, the Fermi level, the maximal electrochemical energy of holes, is situated at a distance below the maximum of the valence band such that:

$$\partial E_p = E_{FP} - E_V = k_B T \, \mathrm{Log}\left(\frac{N_V}{N_A}\right) \qquad [1.3]$$

– In the “N” semiconductor, the Fermi level is situated at a distance below the minimum of the BdC such that:

$$\partial E_n = E_C - E_{FN} = k_B T \, \mathrm{Log}\left(\frac{N_C}{N_D}\right) \qquad [1.4]$$

– If the two semiconductors are part of the same crystalline network and if, at thermodynamic equilibrium (no polarization), the two Fermi levels align:

$$E_G = qV_b + \partial E_n + \partial E_p \qquad [1.5]$$

– A distortion of energy bands occurs. The difference between the minimum of the conduction band on the P side and the minimum of the BdC on the N side corresponds to the variation of the potential energy of the conduction electron. This manifests itself by the appearance of a *voltage barrier*:

$$qV_b = E_G - k_B \, T \, \mathrm{Log}\frac{N_C N_V}{N_A N_D} \ \text{(eV)} \qquad [1.6]$$

V_b: built-in voltage or diffusion potential, thus:

$$qV_b = k_B \, T \, \mathrm{Log}\frac{N_A N_D}{n_i^2} \qquad [1.7]$$

– There is a potential variation, V_{bi}, across a PN junction even if the external polarization is null.

– For a doping of N_A and N_D, being identical, the greater the $E_G \nearrow$, the greater the $V_{bi} \nearrow$ (e.g. SiC, GaN):

- The height of the barrier potential is large in large bandgap semiconductors (e.g. SiC, GaN).

– For a fixed E_G, the greater the $N_A N_D \nearrow$, the greater the $V_b \nearrow$:

- The more the doping in the P and N sections, the larger the potential barrier of the junction.

– When the operating temperature of the junction increases, E_G varies very little, and thus the magnitude of the barrier potential decreases:

- The more the temperature increases, the more the barrier potential of the junction decreases:

$$E'_M = E_M \sqrt{1 - V_j/V_b}. \qquad [1.8]$$

1.4.2. *Directly polarized PN junction*

DEFINITIONS.–

Directly polarizing: the “P” region of the junction is shifted to a positive potential with respect to the “N” region.

We will use V_j to denote the potential difference created at the level of the junction. For direct polarization, the V_j value is positive, where j is the junction.

By applying direct polarization, we can reduce the magnitude of the barrier potential that exists when the junction is non-polarized.

By applying direct polarization, we can reduce the diffusion (retention) electric field. Consequently, the more a junction is directly polarized, the greater the diffusion of the holes in the “P” region towards the “N” region and the diffusion of the electrons of the “N” region towards the “P” regions. The current that is set up is a direct current.

1.4.3. *The reverse polarized junction*

The “P” region of the junction is at a negative potential with respect to the “N” region. V_j, the potential difference created by the external source at the level of the junction, is negative.

– The potential of the “N” region is the same in Figure 1.16. The Fermi level of the “N” region does not change.

– As the energy increases with a diminution in the potential, qV is positive and thus the energy level “rises”.

– When the negative potential difference, V_j, is applied to the junction, the Fermi level of the “P” region increases from the quantity qV_j.

– The magnitude of the potential barrier, which is qV_b without polarization, is now larger and has the value:

$$qV_b = qV_{b0} - qV_j(\text{eV}).$$

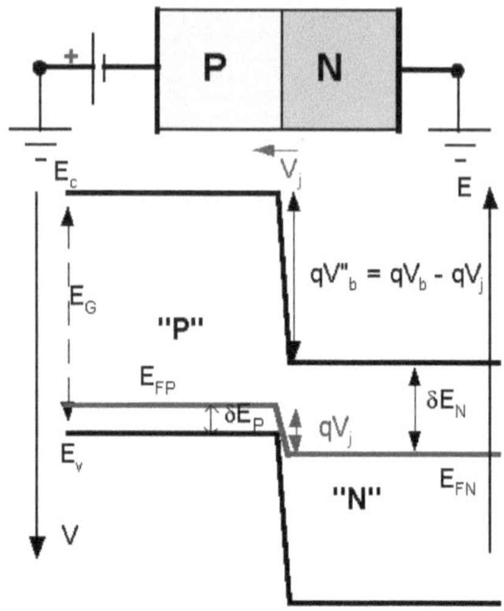

Figure 1.16. *Reversed polarized diode. For a color version of this figure, see www.iste.co.uk/gontrand/nanoelectronics.zip*

By applying reverse polarization, we increase the magnitude of the potential barrier that exists when the junction is non-polarized.

By applying a reverse polarization ($V_j < 0$), we increase the size of the depleted zone.

The more we increase the reverse polarization (V_j becomes more and more negative), the greater the decrease in the junction capacity.

In general:

$$I_D = I_S\left(\exp\frac{qV_j}{k_B T} - 1\right) \qquad [1.9]$$

but as V_j is negative $\exp(qV_j/k_BT)$ tends to 0, the inverse current $I_i = -Is$.

The inverse current traveling across a reverse polarized junction is very weak (10^{-7} A at the ambient), independent of the applied voltage, and varies strongly with temperature (doubling every 8°C):

This comes from the minority carriers (p in the N region and n in the P region). These minority carriers, which "see" the junction directly biased, have a concentration of 10^{10} cm^{-3} in ambient conditions.

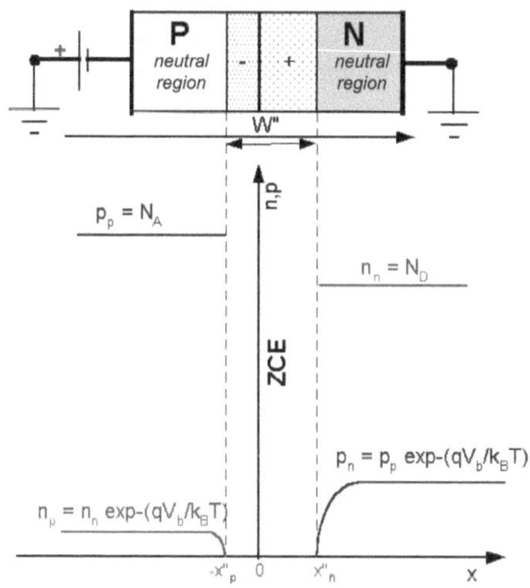

Figure 1.17. *Diode in inverse polarization. For a color version of this figure, see www.iste.co.uk/gontrand/nanoelectronics.zip*

1.4.4. *Polarization of the junction*

Having examined how the junction between two different types of semiconductors is formed, let us look at the behavior of this junction when it is polarized, that is, when the voltage delivered by a continuous supply source is applied to the two zones of the crystal.

The voltage may be applied in both directions, that is, connecting the positive pole of the battery to the N zone and the negative pole to the P zone, or *vice versa*.

In the first case, we say that the junction (that is, the diode) is *reverse* polarized, while in the second case, the diode is *directly* polarized (P to (+), N to (–))

What happens at the level of the PN junction when it is reverse polarized?

Figure 1.20 shows that when the circuit is closed, a part of the free electrons leave the N zone of the crystal and move towards the positive pole of the supply battery.

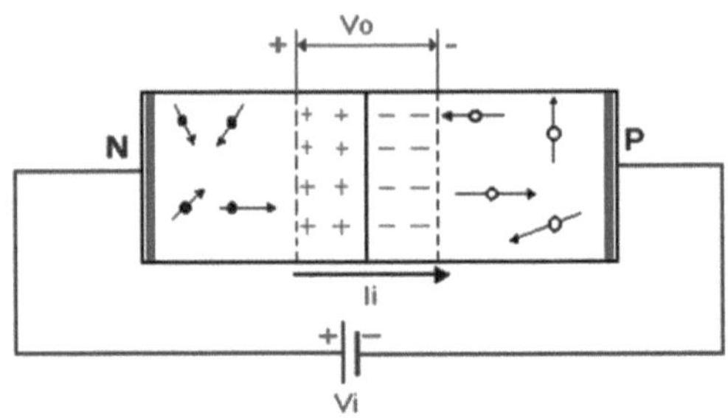

Figure 1.18. *Reverse diode. For a color version of this figure, see www.iste.co.uk/gontrand/nanoelectronics.zip*

At the same time, a certain quantity of electrons emitted by the negative pole of the battery enter the P zone of the crystal, where they cause some of the holes to disappear.

If we now admit that in the P zone, there are no free electrons that can move to the N zone to replace those that were pushed towards the positive pole of the battery, and that in the N zone, there are no holes that may propagate themselves to the P zone in order to replace those that disappeared, we can thus conclude that the movements of the charges between the crystal and the battery, and the battery and the crystal have ceased. In fact, the number of free electrons present in the N zone of the crystal is undeniably very large, but not unlimited. This is the same for the holes present in the P zone of the crystal.

In reality, the movement of the charges and, consequently, the current produced by the battery stops even before the N zone is free of electrons and the P zone is free of holes.

The new potential difference Vo' can thus cancel out the effect of the external tension Vi before all the electrons of the N zone are pushed towards the positive pole of the battery and before all the holes of the P zone disappear.

The voltage Vi, applied to the terminals of the diode, is called the reverse voltage. If we take into account what we discussed earlier, the current circulating in the diode (to whose terminals we have applied a reverse voltage) must cancel itself out very quickly. In reality, the current does not completely cancel out due to the presence of minority charge carriers, that is, the presence of holes in the N zone (with $p << n$) of the crystal and free electrons in the P zone (with $p >> n$). Reminder: we say that the minority carriers "see" the junction directly.

A certain number of minority carriers always succeed in traveling through the junction, thus causing a partial replacement of free electrons in the N zone and of holes in the P zone. We thus observe the presence of a very weak current (~ 10^{-15} A), circulating from the N extremity to the P extremity of the crystal. This current is called the reverse saturation current (Is).

Diode in reverse

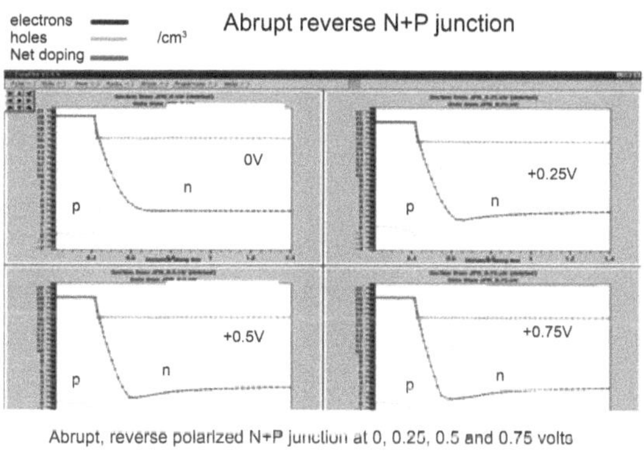

Figure 1.19. *Density of p and n carriers [SIL]. For a color version of this figure, see www.iste.co.uk/gontrand/nanoelectronics.zip*

Let us now look at the converse phenomenon, that is, when the PN junction is *directly* polarized.

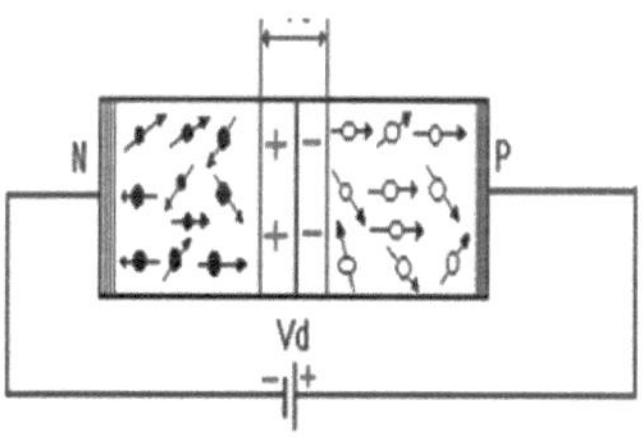

Figure 1.20. *Directly polarized diode. For a color version of this figure, see www.iste.co.uk/gontrand/nanoelectronics.zip*

When the circuit is closed, the electromotor force of the battery sets into motion the free electrons in the N zone and the holes in the P zone, both of which converge towards the junction, within which the electrons fall into the holes, leading to the disappearance of both. However, the free electrons that fall into holes are continually replaced by others coming from the negative pole of the supply source.

Thus, all the holes are replaced by others, which form towards the side of the P zone, towards the positive pole of the battery. The charge flux thus perpetually reproduces itself, forming a continuous current. We can also see this by measuring the direct resistance of the diode.

The continuous current, Id, is called the direct current, and the external voltage, Vd, which leads to the formation of the current Id, is called the direct voltage.

As long as the voltage V_d is lower than or equal to V_o, the current is practically null. This current only exists when the voltage V_d exceeds the value of V_o. This value differs depending on whether the junction is made up of a germanium or silicon crystal: for germanium, this value is usually approximately 0.25 V, while it is approximately 0.6–0.7 V for silicon.

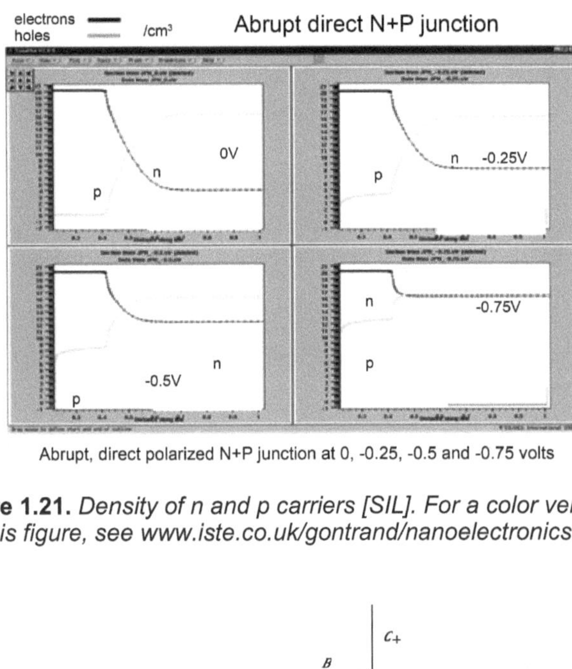

Figure 1.21. *Density of n and p carriers [SIL]. For a color version of this figure, see www.iste.co.uk/gontrand/nanoelectronics.zip*

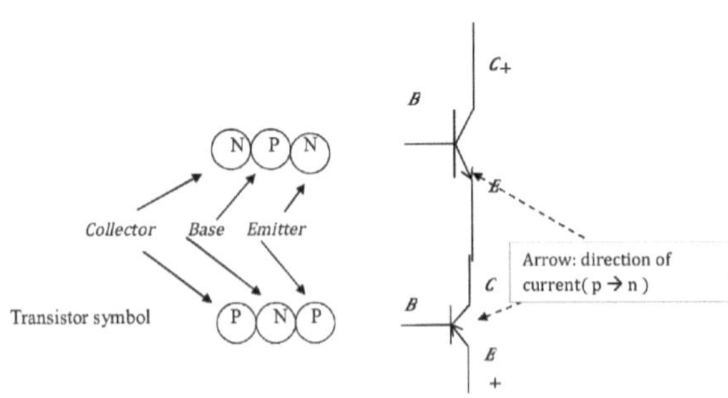

Figure 1.22. *Bipolar schematics*

Technological implementation of an integrated bipolar transistor:

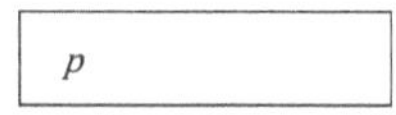

Creation of a substrate (bulk) P

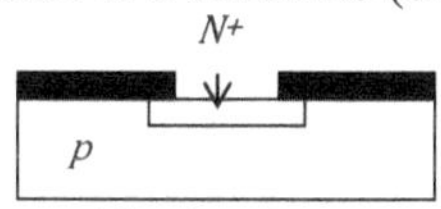

Oxidation – photo-etching of the oxide implementation, then diffusion of the *sole* or *buried layer* of the collector

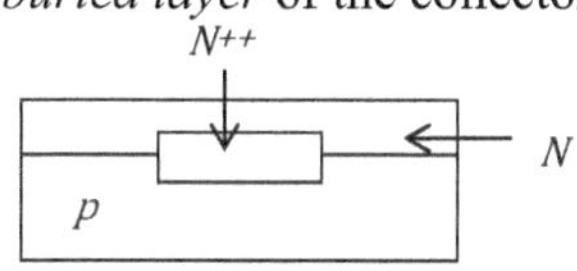

Removing the oxide *Epitaxy* (*epi*: above, *taxis*: classification) of an N^- layer (collector)

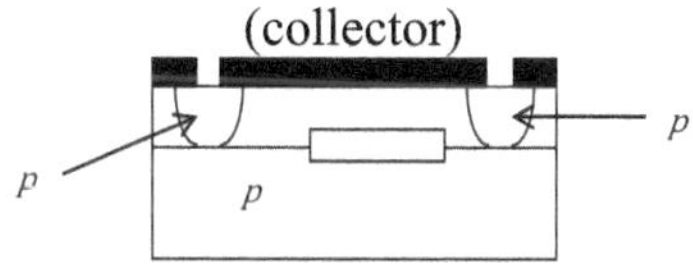

Oxidation – photo-etching of the oxide P diffusion of the channel stops

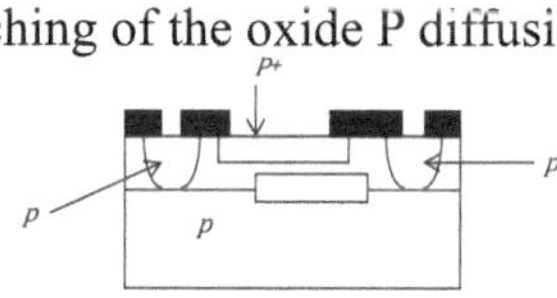

Oxidation – photo-etching of the oxide
P^+ *diffusion of the base* and recharge of the channel stops

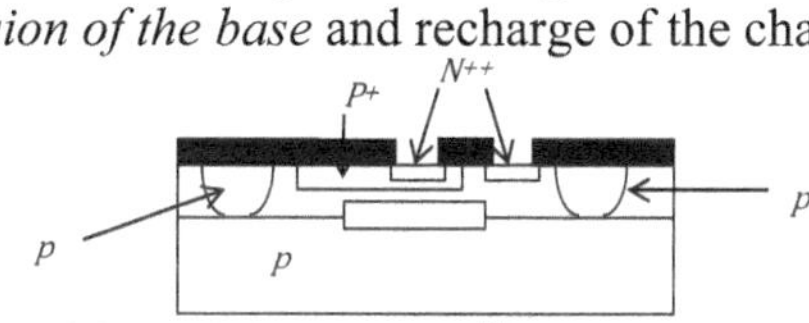

Oxidation – photo-etching of the oxide N^{++} diffusion of the emitter and of the collector contact

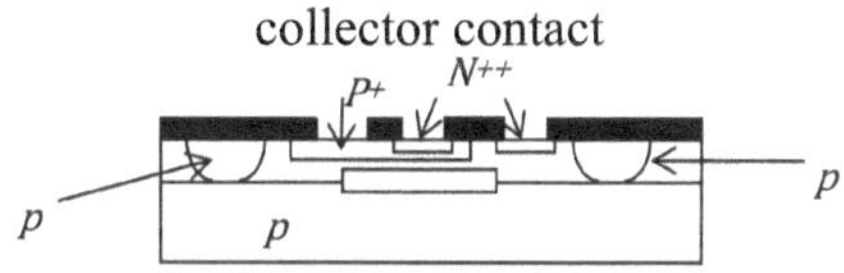

Contacts openings

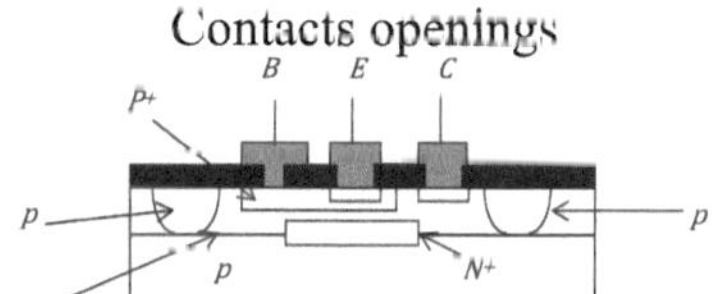

Metallization – etching of the contact. Then, encapsulation – BPSG (Boro-Phosphate Silicon Glass) protection (the glass is ductile, thanks to the boron present)

Figure 1.23. *Working of a bipolar transistor*

1.5. The transistor effect

During the "normal direct" working of the device, the base–emitter junction is directly polarized in this direction, with the other direction (B-C) being the reverse. If we permute the polarization directions of the two junctions, the mode is said to be "normal reverse".

When the base is saturated by the carriers coming from the emitter and the collector, the mode is said to be saturated – the two junctions are directly polarized. When the two junctions are reverse polarized, the mode is said to be blocked.

– For an NPN, the voltage V_{BE} will be positive. In other words, the base is polarized positively with respect to the emitter.

– For a PNP, the voltage V_{BE} will be negative. In other words, the base is polarized negatively with respect to the emitter.

– Flux of the carriers of the emitters towards the base (electrons, in the case of an NPN transistor, the current traveling in the direction opposite to the flux; holes in the case of the PNP transistor, the current traveling in the same direction as the flux).

Additionally, the collector–base junction is strongly reverse polarized (few volts).

– For an NPN, the potential difference V_{CB} will be positive. In other words, the collector is polarized positively with respect to the base.

– For a PNP, the voltage V_{CB} will be negative. In other words, the collector is polarized negatively with respect to the base.

– The carriers in the base (electrons in the case of the NPN transistor and holes in the case of the PNP transistor) are attracted (snapped up) by the collector potential.

– The transistor phenomenon essentially results from the small thickness of the base and the heavy doping of the emitter; the emitter current, I_E, that arrives at the base depends on the direct voltage applied, V_{BE}; the holes in the case of the PNP that arrive at this base are snapped up by the high electric field created by the strongly negative voltage V_{CB} (in the case of the PNP). A large part of these holes travel across the narrow base to penetrate the collector and form the current I_C. This collector current is thus not very dependent on V_{CB} so long as this polarization is sufficient.

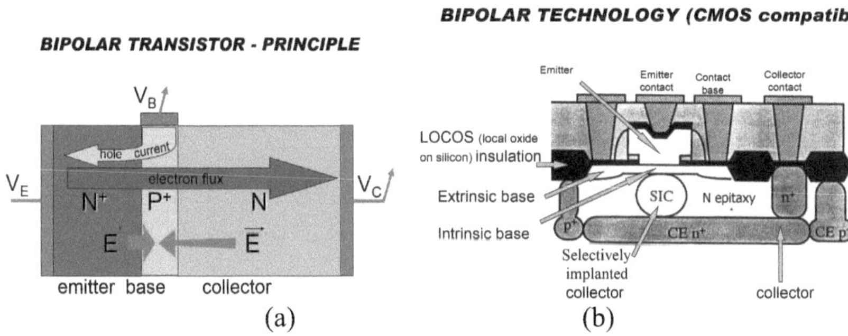

Figure 1.24. *(a) The NPN bipolar transistor; (b) BiCMOS technology. For a color version of this figure, see www.iste.co.uk/gontrand/nanoelectronics.zip*

1.5.1. *Controlling the current gain*

Three determinant factors must be considered to calculate the current gain:

a) injection efficiency: γ;

b) the transport factor: B_T;

c) the multiplication factor at the base–collector (in reverse) junction crossing: M.

However, be careful, Ic is no longer controlled by Ib!

The current gain, α, is the product of these three factors: $\alpha = \gamma B_{TM}$.

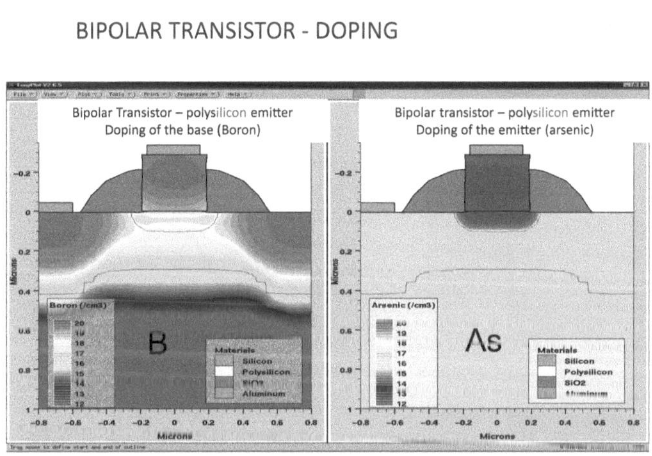

Figure 1.25. *SILVACO simulations (finite elements). For a color version of this figure, see www.iste.co.uk/gontrand/nanoelectronics.zip*

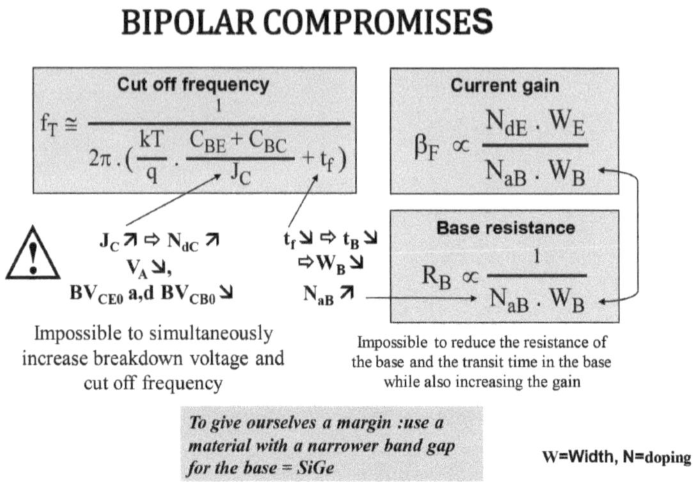

Figure 1.26. *Relations between process and electrical parameters concerning the bipolar transistor trade offs*

1.6. MOSFET (Metal Oxide Semiconductor Field-Effect Transistor)

1.6.1. *Base structure*

The MOSFET is made up of two heavily doped zones/reservoirs called *source* and *drain* (semiconducting regions with the same type of doping), each connected to their respective electrode. Source and drain are implemented in a region (sometimes directly in the substrate) of the opposite type.

– A command electrode (*gate*), which overhangs the channel zone.

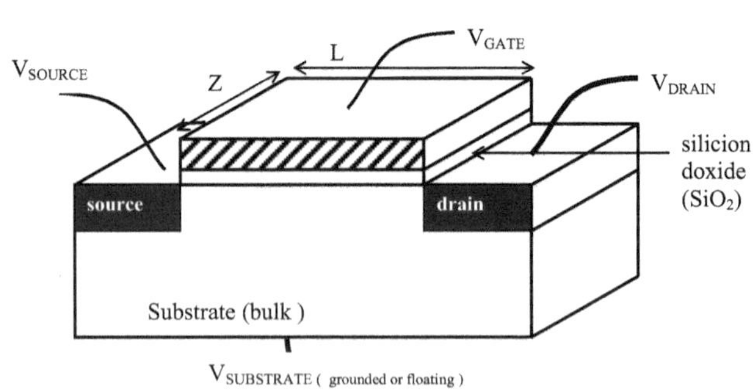

Figure 1.27. *Structure of the MOS transistor*

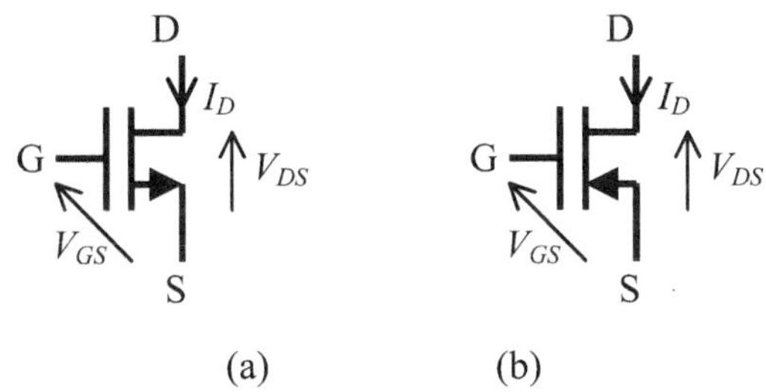

Figure 1.28. *(a) NMOS symbol, (b) PMOS symbol*

The current in the MOS transistors is a *unipolar current* of majority carriers.

The arrow in the source–gate branch indicates the direction of the current (+ ---- –).

– Electrons in the case of an NMOS transistor (source and drain are N-type).

– Holes in the case of a PMOS transistor (source and drain are P-type).

1.6.2. *Functioning principle*

1) The gate (G) makes it possible to control the current between the two electrodes source (S) and drain (D).

2) The control is exerted via the applied potential between the gate and the source. This system is made up of an MOS (Metal–Oxide–Silicon) capacity and allows commands with very low consumption of current. We talk of isolated gate transistors (see displacement current: variation of the flux of the electrical field versus time).

3) We can compare the drain–source system to a resistance whose value varies between R_{on} (min value: closed circuit) and R_{off}.

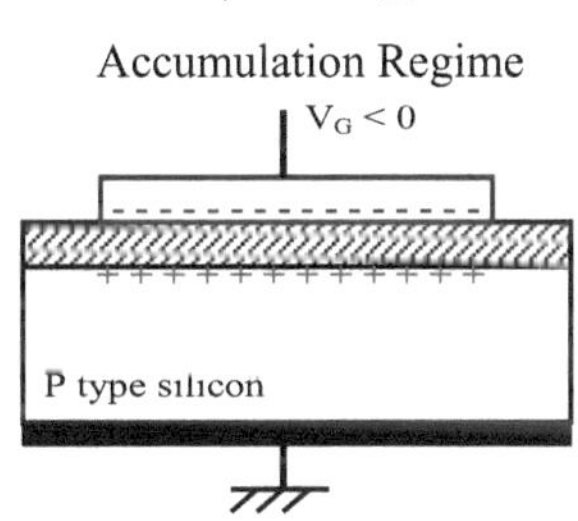

Figure 1.29. *Accumulation of majority carriers on the surface of the silicon (enrichment regime). For a color version of this figure, see www.iste.co.uk/gontrand/nanoelectronics.zip*

Depletion regime (desertion)

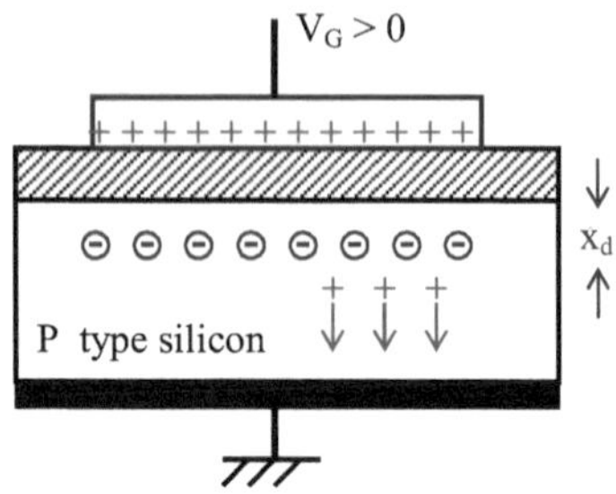

Figure 1.30. *MOS physical functioning. For a color version of this figure, see www.iste.co.uk/gontrand/nanoelectronics.zip*

Depletion regime or impoverishment of the silicon surface:

– *majority carriers are pushed deep inside*;

– *negative fixed ions remain.*

Inversion regime (mostly used vs. "the enrichment regime")

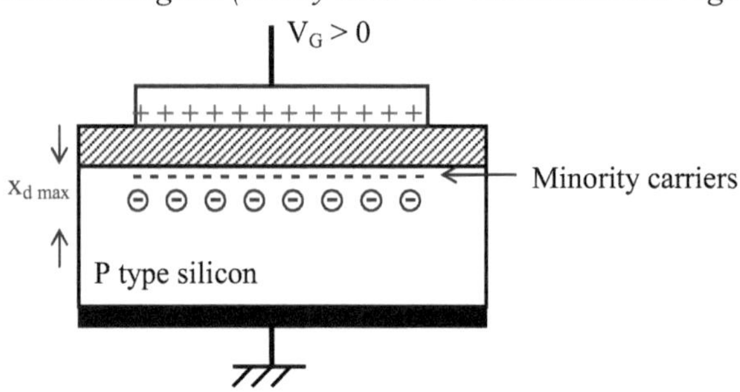

Minority carriers= electrons

Figure 1.31. *Working of the NMOS. For a color version of this figure, see www.iste.co.uk/gontrand/nanoelectronics.zip*

1.6.3. *Electric model for the MOS capacitor and C(V) curve*

Electric model of the MOS capacitor (see command gate – analogy with a tap and its joint: oxide Figures 1.31 and 1.32): the MOS capacitor is the equivalent of two capacitors connected in series: the oxide capacitor (C_{ox}) and the semiconductor capacitor ($\mathbf{C_{SC}}$).

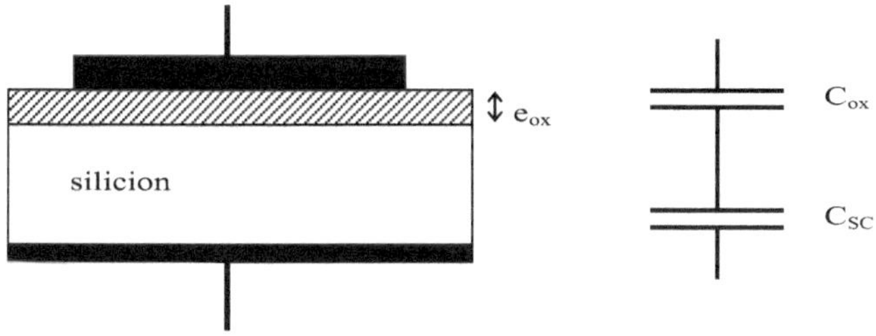

Figure 1.32. *MOS "capa"*

2D Meshing

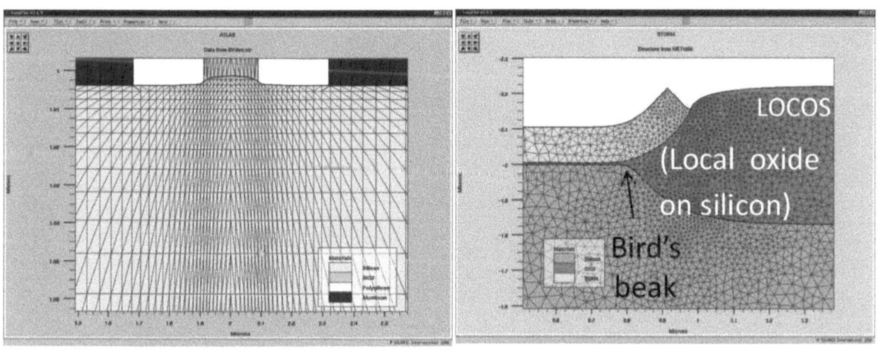

Regular mesh for finite differences

Irregular mesh for finite elements

Figure 1.33. *MOS meshes. For a color version of this figure, see www.iste.co.uk/gontrand/nanoelectronics.zip*

2

Modeling of Diffusion Processes

SUMMARY.– In this chapter, we present essential information on the manufacture of integrated electronic devices. In the present case, we will emphasize the diffusion of dopants, which is a crucial step. We will thus study the diffusion equation and how it is used in the context of modeling in the manufacture of microelectronic devices. However, from the nanoelectronic point of view, we will also discuss some theories of irreversible processes, the random movements of dopants and the diverse types of dopants. Some very simple examples from informatics codes conclude the chapter, allowing the reader to carry out some numerical experiments, as well as helping to extend these examples in different directions.

2.1. Introduction

Diffusion processes are the most crucial steps in the manufacture of integrated circuits. They are underpinned by the basic equation: the diffusion equation, also known as the heat equation.

In the 1970s, physicists and technologists were led to work closely on bidimensional problems (e.g. lateral diffusions, diffusion accelerated by oxidation or due to the inevitable presence of other dopants); indeed, they worked so closely that their roles were hard to distinguish.

We must also keep in mind that the physics of the occasional defects, metallic defects, for example, in the context of semiconductors are far from being taken into account in commercial simulators: miniaturization may aggravate these problems.

Since the 1980s, ion implantation has replaced initial diffusions: the mean depth of the implanted doping agent is well controlled, and the subsequent rapid thermal annealing (RTA) makes it possible to adjust post-implantation diffusions (drive-in).

The diffusion equation – for the dopants, here – is one of the fundamental equations in physics and involves the dreaded Laplacian operator!

Mathematicians at the highest level are, even today, trying to justify these in the most in-depth ways possible. Another example is fluids (Euler and Navier–Stokes); would it be possible to derive these equations from microscopic equations, related to Newtonian mechanics?

Let us return to the diffusion equation: for several decades, this equation has been related to microscopic random walk-type processes, or to Brownian motion, typically random (see Markovian interaction; in brief, the future of the particle depends on the present and not on the past), knowing that breakthroughs in mathematics are now emerging.

Microscopic equations of the Newtonian type may be applied to all of the particles of the system under consideration (e.g. electrons in a gaseous semiconductor). However, their number is highly prohibitive.

We thus move on to the mesoscopic level of kinetic equations (Boltzmann, first half of the 19th Century): we define the densities of carriers (of electricity) or particles in the phases SPACE (SPACE of positions and speeds or wave vectors). In these equations, we can see partial derivatives with respect to time. The third and final level is that of macroscopic equations: these equations (e.g. Navier–Stokes) depend on time and speed, but not on position.

Moving from microscopic to mesoscopic, and finally to macroscopic, makes these models less precise (see "information loss") but easier to handle.

For example, going from meso to pico is the same as going beyond nanoelectronics to picoelectronics (between the atom and its nucleus, the field of electrons; as gas, thus microscopic).

In our case, we can write that the density of Brownian particles obeys the heat equation, which models thermal conductivity.

In addition, it has been demonstrated [LAN 75] that a gas of particles obeys the kinetics of the Boltzmann equation, even if this equation is initially highly out of equilibrium (even locally), although confined to one site, or if the particles are homokinetic. However, this is for highly diluted gases and for a short time (see Knudsen's criterion: the ratio between nominal time and that of the mean free path (of relaxation time between two shocks)). This theorem seems to state that no particle system can "go back in time", nor return to a state out of equilibrium, nor return to fluidic regimes.

In the case of diffusion, if we consider a particle whose position or speed has been varied very little (to remain in the linear regime), then if we follow the process of collisions between particles for long enough, this latter tending "slowly" towards infinity and if repeated shocks between two particles are rare, the limit is Brownian motion (this result is known to those who use the Monte Carlo methods – statistics – that are applied to the resolution of the Boltzmann equation (see Chapter 3), where the elastic–acoustic phonon interactions with electron gases (or holes) are predominant and there is a very weak electric field). We also know that Brownian motion and diffusion are very closely related: what about the possible relation between the Boltzmann equation and the diffusion equation? The Laplacian considers real space, while the collision term (see "Boltzmann") acts in the space of speed (or wave vectors), and the union of these two spaces forms the space of phases. For example, in the static case, the concentration becomes constant or linear in real space, whereas, on the contrary, in the collision term – where the input and the output in a state *k* is null – the inputs and outputs in this state arrive at an equilibrium; for example, some carriers take large speeds, while a large number of carriers may acquire a small speed. This equals a constant flux or current:

$$n\,v = cste.$$

Thus, if we start from hard balls with elastic shocks – without gain or loss of energy – with the shocks obeying Newton's laws of mechanics, we arrive at a random movement without memory.

Other gas kinetics show that via the fundamental theory of dynamics, we may arrive at another problem, one that is *a priori* much bigger – that the micro–meso shift will be moving from the microscopic to the macroscopic, with or without going through the mesoscopic for a long time – much higher than the mean free time. Another difficulty is shifting to less diluted media: a greater number of collisions per unit time.

2.2. Phenomenological equations of diffusion [MAT 96, ADD 66, BOU 03]

When all dopants in the silicon are ionized in ambient conditions, there is a concentration gradient of majority free carriers in the semiconductor substance. As its current density is null (a thermodynamic equation – in practice, no electric polarization), the diffusion effect of carriers is compensated by the electric field effect E.

Considering a material that consists of donor atoms at a concentration $N_D(x)$, the density of the total electron current can be expressed (e.g. in 1D) by considering conduction and diffusion, through:

$$J_n = qD_b \frac{\partial n}{\partial x} + qn\mu_b E \qquad [2.1]$$

In this equation, we can initially hypothesize that the concentration of electrons at all points is equal to the concentration of ionized dopant atoms. At thermodynamic equilibrium, the total current density is null, Jn=0; Hence:

$$E = -V_T \frac{1}{n}\frac{\partial n}{\partial x}, with\ V_T = \frac{kT}{q} \qquad [2.2]$$

Moreover, if C(x,t) represents $N_D(x)$ and n, in particular under ambient conditions, in the silicon,

$$\frac{\partial n}{\partial x} = \frac{\partial C(x,t)}{\partial x} = \frac{\partial N_D(x)}{\partial x} \qquad [2.3]$$

$$E = V_T \frac{\partial}{\partial x}(ln \frac{N_D(x)}{n_i}) \qquad [2.4]$$

As the displacement of ionized atoms is equivalent to a current density, the flux of dopant atoms can be expressed as:

$$J(x) = -qD \ \frac{\partial}{\partial x}(N_D(x)) - qz\frac{N_D(x)}{V_T}E(x) \qquad [2.5]$$

with Z being the charge of the dopant (1, 2, …)

$$J(x) = -qD \ h\frac{\partial}{\partial x}(N_D(x)) \qquad [2.6]$$

with h being the exaltation factor of the field, which can be thus written as:

$$h = 1 + zN_D(x) \ \frac{\partial}{\partial N_D}(ln \frac{N_D(x)}{n_i}) \qquad [2.7]$$

or again:

– if $N_D << n_i$ at diffusion temperature, then h=1;

– if $N_D >> n_i$ at diffusion temperature, then h=2 (maximal value).

We can see that the intrinsic concentration, n_i, strongly depends on the temperature. Indeed, n_i is thermally activated and can be expressed as:

$$n_i = N_D N_V \exp(-\frac{E_G}{2kT}) \qquad [2.8]$$

For silicon, at ambient temperature (300 K), n_i is of the order of 10^{10} cm^{-3}, and E_G of the order of 1.12 eV. At 1,027°C, that is, at 1,300 K, n_i has a value of about $2.5.10^{17}$ cm^{-3}, which means that its order of magnitude is completely modified. This implies that in regions with heavy doping, diffusion will be higher than in regions with weak doping due to the induced electric field. This results in the profile being modified, for example, Gaussian, to have a more abrupt profile.

Until the 1980s, operation at high temperatures had been lasting for tens of seconds or minutes. This then came down to several seconds using RTA, up to the point where it is often no more than a flash today! This reduces the redistribution or diffusion of doping agents. The distribution curve for doping species will thus be modified due to the thermal diffusion that takes place: under the effect of the temperature, doping impurities will move from sites of high concentration to those of lower concentrations.

In Figure 2.1, we present a schematic depiction of principal diffusion mechanisms: we will explain them later, for the purpose of microscopic studies, going towards a drastic reduction in the size of devices.

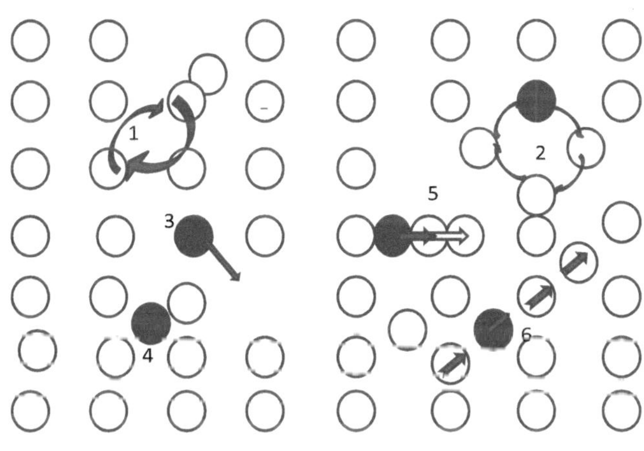

Figure 2.1. *Diagram of principal diffusion mechanisms: (1) simple exchange; (2) cyclic exchange; (3) vacancy-based; (4) direct interstitial; (5) indirect interstitial; (6) "crowdion" [PHI 91]. For a color version of this figure, see www.iste.co.uk/gontrand/nanoelectronics.zip*

To control the electric properties of an electronic device such as a transistor, we must thus be able to control, as precisely as possible, the distribution of the doping species in the silicon, via the diffusion steps.

It is therefore necessary to understand and model elementary mechanisms that govern the codiffusion of dopants. This will be the objective of the following sections.

2.2.1. *Fick's first law*

Fick's first law states that the diffusion flux is proportional to the concentration.

Let us consider an element A "dissolved" in a crystalline matrix. At low temperatures, the system is fixed and the impurity is immobile. However, if the temperature is sufficiently elevated, under the effect of thermal agitation, the atoms of A are mobile and can move around within the matrix. They are thus subject to the diffusion phenomenon: a non-uniformity of thermodynamic variables (a variation in the concentration of A, most often) brings about a transport of material (displacement of A atoms) within the solid. Thus, diffusion is a non-reversible process whose "motor" is the gradient of the chemical potential (or variation in the chemical potential within the matrix of the substrate) and whose ultimate result is making this chemical potential uniform. The general description of this process is only possible using the thermodynamic theory for irreversible processes.

The gradient of the chemical potential corresponds to the variation in free enthalpy of the system when the number of moles of the species in space is varied (all other things remaining equal):

$$\vec{\nabla}\mu_i = \vec{\nabla}\left(\frac{\partial G}{\partial n_i}\right)_{P,T,n_{j(j\neq i)}} \qquad [2.9]$$

DEFINITION: FREE ENTHALPY.–

The free enthalpy of a system is the function of the state *G* (F: Gibbs) defined by:

$$G = A + PV = U - TS + PV = H - TS \qquad [2.10a]$$

with A being the free energy and U being the energy.

During an elementary, reversible translation (thus starting from a state of equilibrium), we have:

$$dG = VdP - SdT \qquad [2.10b]$$

G is thus a function of temperature and pressure.

Finally, we can show that during a real isothermal, isobaric transformation of a system, its free enthalpy can only decrease until it reaches a state of equilibrium, with the equality:

$$\Delta G=0 \quad [2.10c]$$

corresponding to a reversible transformation.

When a system changes at fixed temperature and pressure, its free enthalpy decreases until it reaches a state of equilibrium.

However, in the simplest case of a single diffusing species subject only to its concentration gradient, the phenomenon can be very simply described using Fick's first law, according to which the flux of an impurity J (number of atoms traversing a unit of surface in a unit of time) is proportional to the concentration gradient of the impurity, C.

$$\vec{J} = -D\vec{\nabla} C \quad [2.11]$$

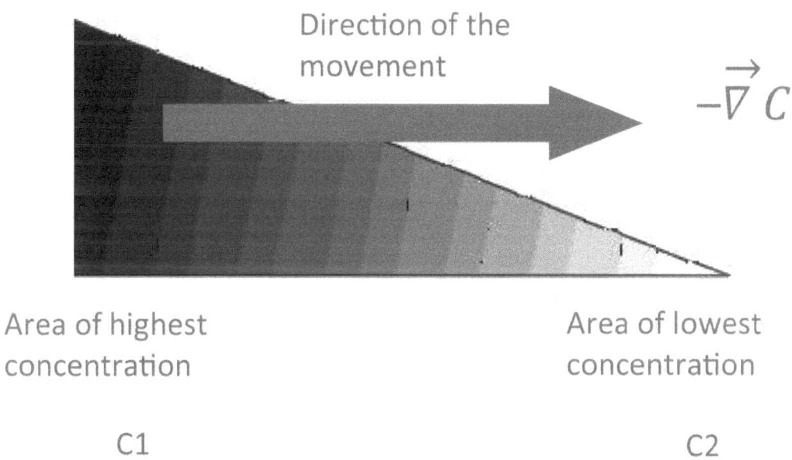

Figure 2.2. *A concentration gradient sets up the flux. For a color version of this figure, see www.iste.co.uk/gontrand/nanoelectronics.zip*

The "–" sign in the preceding equation simply shows the fact that the diffusion takes place from zones of high concentration towards zones of low concentration, tending to homogenize the concentrations. The proportionality coefficient, D, is the diffusion coefficient or diffusivity; it depends on the temperature. As the flux is generally expressed in atoms per square centimeter and per second ($cm^{-2}s^{-1}$) and the concentration in number of atoms per cubic centimeter (cm^{-3}), in practice, D is expressed in cm^2s^{-1} (spread over a surface over time).

Flux is a vector quantity where, in the most general case (anisotropic medium), D is a tensor. Thus, flux is not necessarily parallel to the concentration gradient. This may occur in non-cubic materials or materials that are deformed under the effect of a stress, for example.

However, in the case of silicon, which is a cubic material and thus isotropic (properties of the medium are identical in all the directions observed), when deformation is absent, the deformation coefficient is scalar. The flux given by Fick's first equation is thus parallel to the concentration gradient. This is why, in the following sections, we will consider only the unidimensional case for which Fick's first law is reduced to (here and below in 1D):

$$J_x = -D_x \frac{\partial C}{\partial x} \qquad [2.12]$$

2.2.2. *Generalized flux*

In the previous sections, we assumed that the diffusant impurity was only subject to its concentration gradient. Let us now consider the more general case where the impurity is subject to mortice forces of diverse origins (concentration gradient, stress field, electric field, etc.), and that under the effect of these forces (and their interactions with the crystalline network), the impurity acquires a mean stationary speed, u. We can thus show that the flux of impurities is given by the equation:

$$\vec{J} = \vec{u} C \qquad [2.13]$$

The general problem is the same as identifying the nature of the different forces and establishing the relation between speed and force. If the driving force is only due to the concentration gradient, the thermodynamics of irreversible processes makes it possible to show that the driving force corresponds to the partial mixing entropy gradient. In the general case of a set of driving forces, admitting that the average speed of the impurity is the sum of the elementary speed induced by each of the driving forces taken separately (see linearity), it is convenient to write the impurity flux in the form:

$$J = -D \frac{\partial C}{\partial x} + uC \qquad [2.14]$$

In the previous equation, the first term (called the "Fickian") represents the contribution of the concentration gradient alone and the second term (the "Nernst" or "drag term") represents the contribution of all the other driving forces. An important case is that of a charged species in the presence of an electric field. In these conditions, the drift speed is given by the relation:

$$\vec{u} = \pm \mu \vec{E} \quad [2.15]$$

where μ is the mobility and E is the electric field. The sign ± depends on the sign of the impurity charge. In these conditions, the flux can be simply written as:

$$J = -D\frac{\partial C}{\partial x} \pm \mu EC \quad [2.16]$$

As we will see further on, this expression of flux, which is well known in the theory of transport of electrical carriers, is also used for the problem of the diffusion of dopants in a semiconductor.

2.2.3. *Fick's second law*

Established for a conservative system, in other words, where there is neither the loss nor creation of atoms of the diffusant species, Fick's second law makes it possible to describe the temporal evolution of the concentration in the diffusant species at a given point. A simple balancing of the substance leads to the following conservation (or continuity) equation (recall):

$$\frac{\partial C}{\partial t} = -\frac{\partial J}{\partial x} \quad [2.17]$$

From this, we obtain:

$$\frac{\partial C}{\partial t} = -\frac{\partial}{\partial x}\left(D\frac{\partial C}{\partial x}\right) \quad [2.18]$$

If D is considered to be constant, that is, independent of the concentration, and thus independent of x and t, in the hypothesis of a diffusion subject only to the effect of a concentration gradient, equation [2.18] gives (recall):

$$\frac{\partial C}{\partial t} = D\frac{\partial^2 C}{\partial x^2} \quad [2.19]$$

As shown already, the Laplacian (see second derivative) indicates that for a dip in the doping profile (concavity/positive curve), C increases as a function of time (the dip is filled up) and, reciprocally, a hump (negative concavity) in the profile will flatten out over time.

For a species that is subject to additional drag forces, equation [2.14] with D constant leads to:

$$\frac{\partial c}{\partial t} = D\,\frac{\partial^2 c}{\partial x^2} - u\frac{\partial c}{\partial x} - c\frac{\partial u}{\partial x} \qquad [2.20]$$

When the system is not conservative, for example in the case of a chemical reaction leading to the formation or elimination of impurities, or trapping/freeing impurities, and so on, these equations are modified by the addition of terms for generation or recombining, which depend on the mechanisms implemented. Equation [2.20] (Fick's second equation) is therefore valid only in very restrictive conditions that rarely come together for diffusion problems in semiconductors. Indeed, the simplifying hypotheses used are almost never seen in practice. For instance, here we consider a coefficient D to be constant, while in reality, it depends on the temperature and concentration of the diffusant species.

Equations [2.18] and [2.20] make it possible to calculate C(x,t), but as they are equations with partial derivatives, the first order for t and the second order for x, they do not have unique solutions. The appropriate solution for a given situation is determined by the knowledge of an initial condition and two limiting conditions that translate the experimental reality. In general, it is not possible to analytically resolve these equations, which implies that we turn to numerical solutions on a computer.

We can, however, obtain simple analytical solutions for particular initial and limiting conditions. These conditions are generally imposed from an experimental point of view, in order to determine the value of diffusion coefficients by comparing between the measured profiles and these simple laws.

2.2.4. *Codiffusion(s)*

In the field of microelectronics and, by extension, nanoelectronics, silicon is by far the most widely used material. Even though there are alternatives, for instance AsGa (gallium arsenide), silicon is found in abundance and is therefore less expensive, giving it a huge considerable advantage over its competitors (indeed, it is the second most abundant element in the earth's crust, behind oxygen). In addition, it is possible to refine it quite easily up to 98%.

It does, however, have some faults, such as getting oxidized in an insulating species (silicon dioxide) and, above all, being a very poor conductor at constant temperature, almost an insulator, its conductivity being of the order of a hundredth of a billionth that of copper, which is not ideal in the field in which it will be used. To mitigate this problem, we turn to "doping" procedures.

Doping a semiconductor involves diffusing or, nowadays, injecting and implanting doping elements, also called impurities, into the material whose conductivity we wish to improve. This generally concerns cations (electron acceptors) or anions (electron donors), which we inject into two different zones in the material, thus creating what is called a P/N junction, namely a contact surface between a zone with excess electrons and a zone with a deficit of electrons.

While the diffusion mechanisms for species within a crystalline material are relatively well known, there are still questions related to the precise physical modeling of the diffusion of dopants such as arsenic, boron and others in silicon, mainly for low thermal budgets, high doses or implantation.

2.3. Calculation of flux and the internal electric field effect

The flux of an impurity "e" in an isotopic medium containing "j" impurities, knowing that the only parameters are driving force, is given by:

$$\overrightarrow{J_e} = -D_e\,\vec{\nabla} C_e + z_e \mu_e C_e \vec{E} \qquad [2.21]$$

where:

D_e is the diffusion coefficient of the impurity;

C_e is the concentration of the impurity;

μ_e is the mobility of the impurity e;

z_e is the charge on the impurity, positive for donors and negative for acceptors;

$\vec{E}$ is the internal electric field;

∇ is the gradient operator.

If Ψ is the electrostatic potential in the structure, we then have:

$$\vec{E} = -\vec{\nabla}\Psi \qquad [2.22]$$

$$\frac{\partial Ce}{\partial x} = \vec{\nabla}.\overrightarrow{J_e} \qquad [2.23]$$

$$\frac{De}{\mu e} = \frac{kT}{q} \qquad [2.24]$$

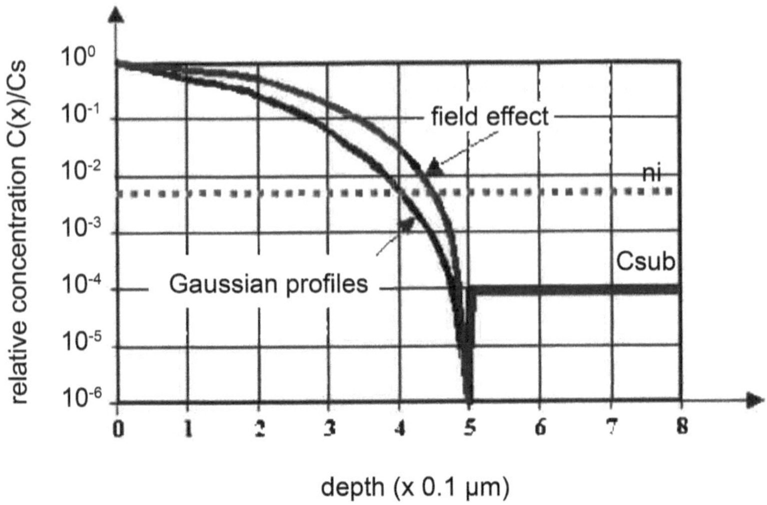

Figure 2.3. *Diffusion accelerated by the internal electric field*

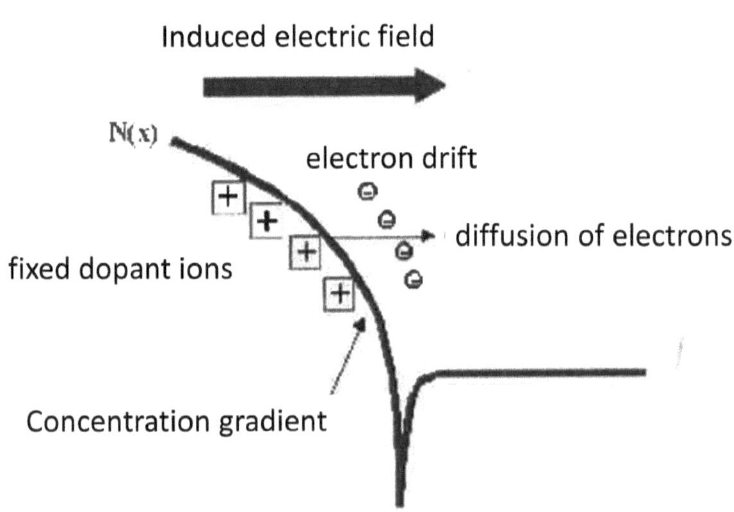

Figure 2.4. *An electron concentration gradient is associated with the dopant concentration gradient of type n (e.g. As^+). An electric field is set up by the diffusion of electrons. This field will also act on the diffusion of ions*

Taking into account the existing difference between the time constants relative to the diffusion of impurities (a few seconds or, sometimes, minutes) and charge carriers (a few microseconds), the following hypotheses can be stated:

a) neutrality of charges;

b) the population of electrons and holes are at thermodynamic equilibrium and obey Boltzmann statistics.

If “n” denotes the concentration (or density) of electrons and “p” denotes the density of holes, electrical neutrality at any point – hypothesis (a) – leads us to write:

$$n^2-(\sum z_j C_j)n-n_i^2=o \qquad [2.25a]$$

and

$$p.n = n_i^2 \qquad [2.25b]$$

$$n = \tfrac{1}{2}(\sum z_j C_j) + \sqrt{(\sum z_j C_j)^2 + 4n_i^2)} \qquad [2.26]$$

and thus

$$p = \tfrac{1}{2}(-\sum z_j C_j) + \sqrt{(\sum z_j C_j)^2 + 4n_i^2)} \qquad [2.27]$$

The sum on “j” is made on all the impurities of the structure. We now assume that all the impurities are ionized during the annealing. If “n_i” denotes the intrinsic electron concentration at the considered temperature, hypothesis (b) gives:

$$p.n=ni^2$$

The “gap” of the semiconductor and n_i depend in fact on doping, as this is very large; here again, for the time being, we neglect this factor.

Knowing that:

$$n=ni*(\exp-(q\Psi/kT)) \qquad [2.28]$$

From [2.22] and [2.28], the expression for flux given by [2.21] becomes:

$$\vec{J}e = -D_e\,(\vec{\nabla} C_e + z_e Ce\vec{\nabla}\log(\tfrac{n}{n_i})) \qquad [2.29]$$

$$\vec{J}e = -D_e\,(\vec{\nabla}C_e + z_e Ce\frac{\Sigma z_j \vec{\nabla} C_j}{\sqrt{\Sigma(z_j C_j)^2 + 4n_i^2}}) \qquad [2.30]$$

Application: in the case of arsenic:

$$\overrightarrow{J_{As}} = -h.\,D_{As}(\,\vec{\nabla} C_{As}) \qquad [2.31]$$

$$h = 1 + C_{As}\frac{1}{\sqrt{1+\left(\frac{C_{As}}{2n_i}\right)^2}}) \qquad [2.32]$$

where h is the exaltation (enhancement) factor of the diffusion.

Then, for strong concentrations ($C_{As}>>n_i$), h=2; the electric field thus contributes to diffusion to the same extent that the concentration of diffusant species does.

2.4. Influence of different vacancies on the diffusion coefficient [WAT 64, MAT 91, HU 69, CHO 85, MIC 87]

Let us now look at how the influence of different vacancies is taken into account for the definition of the diffusion coefficient D. These vacancies may present themselves in different electric states: neutral vacancies, positively charged, negatively charged and so on. The diffusion coefficient, D, is thus the sum of various diffusion coefficients, taking into account the diffusion of the impurity with each of these vacancies:

$$D = D^o + D^-.(n/n_i) + D^{=}\,.(n/n_i)^2 + D+\,.(n/n_i) + ... \qquad [2.33]$$

By normalizing these different diffusion coefficients with respect to the intrinsic diffusion coefficient D_i, which corresponds to the case of diffusion at low concentration ($Ce<<n_i$), and by assuming that the diffusion only takes place for donors (arsenic, for instance), only through the intermediary between neutral and negatively charged vacancies, it follows that:

$$Di = D_i^o + D_i^- \qquad [2.34]$$

$$D_d = D_i^o + D^-\,(n/n_i) = (D_i^o + D_i^-\,[D_i^o + D^-\,(n/n_i)\,/\,(Di'' + Df) \qquad [2.35]$$

Hence:

$$D_d = D_{id}\,[1 + ß_d(n/n_i)]/[1 + ß_d] \qquad [2.36]$$

where $ß_d = (D^-/D^0)$, with d being the donor impurity (e.g. P, As, Sb) having $z_e > 0$.

Similarly, for acceptor impurities (Ze<0), assuming that diffusion takes place through the intermediary between neutral and positively charged vacancies:

$$D_a = D_{ia}\,[1 + ß_a(n/n_{i)}]/[1 + ß_a) \qquad [2.37]$$

where ß is the factor, representing the ratio of diffusivity induced by the vacancies charged once to the diffusivity due to neutral vacancies.

Going back to the general equation for flux [2.21], we obtain:

$$\vec{F_e} = -D_e \frac{1 + ß_e\left(\frac{n}{n_i}\right)}{1 + ß_e} (\vec{\nabla} C_e + z_e Ce \frac{\sum z_j \nabla C_j}{\sqrt{\sum (z_j C_{j)})^2 + 4n_i^2}}) \qquad [2.38]$$

with D_{ie} being the intrinsic diffusion coefficient of the impurity "e";

- f = n/n_i for a donor impurity;

– f = $p/n_i = n_i/n$ for an acceptor impurity.

2.5. Calculation of the electric field in the case of arsenic

Ion and electron flux can be expressed by the following two equations, using [2.21]:

$$\overrightarrow{J_{As}} = -D_{As}\,\vec{\nabla} C_{As} + \mu_{As} \vec{E} C_{As} \qquad [2.39]$$

and

$$\vec{J_n} = -D_n\,\vec{\nabla} n + \mu_n \overrightarrow{E_n}) \qquad [2.40]$$

$$\overrightarrow{J_{As}} = -\vec{J_n} \;=>\; \vec{E} = \frac{D_{As}\vec{\nabla} C_{As} - D_n \vec{\nabla} n}{\mu_n n + C_{As}} \qquad [2.41]$$

The value of $\vec{\nabla} n$ can be determined using [2.26]:

$$\vec{\nabla} n = [1 + \frac{C_{As}}{\sqrt{(2n_i)^2 + (C_{As})^2}}]\vec{\nabla} CAs \qquad [2.42]$$

By substituting the value of $\vec{\nabla}$n, and using Einstein's relation, relation [2.41] becomes:

$$\vec{E} = \frac{kT}{q} \left[\frac{\left[D_{As} - D_n \frac{n}{\sqrt{(2n_i)^2 + (C_{As})^2}}\right]}{\left[D_n \frac{n}{C_{As}} + D_{As}\right] C_{As}}\right] . \vec{\nabla} C_{As} \qquad [2.43]$$

Knowing that:

$$n/C_{As} = 1 + (p/C_{As}) \qquad [2.44a]$$

and

$$p/n = (n_i/n)2 \qquad [2.44b]$$

we finally obtain:

$$\vec{E} = \frac{kT}{qC_{As}} \left[\frac{D_{As} - D_n[1 + (\frac{n_i}{n})^2]^{-1}}{D_{As} + D_n + D_n\left(1 + \frac{n_i^2}{nC_{As}}\right)}\right] \vec{\nabla} C_{As} \qquad [2.45]$$

Similarly, Kurt and Alexis [KUR 61] obtain Ë in the case of an acceptor (e.g. boron):

$$\vec{E} = \frac{kT}{qC_B} \left[\frac{D_B - D_p[1 + (\frac{n_i}{p})^2]^{-1}}{D_B + D_p + D_p\left(1 + \frac{n_i^2}{np}\right)}\right] \vec{\nabla} C_B \qquad [2.46]$$

Using [2.39] and [2.45], we obtain the expression for flux:

$$\overrightarrow{J_{As}} = -D_{As} \left[1 - \frac{D_{As} - D_n[1 + (\frac{n_i}{n})^2]^{-1}}{[D_{As} + D_n + D_n\left(1 + \frac{n_i^2}{nC_{As}}\right)]} \vec{\nabla} C_{As}\right] \qquad [2.47]$$

Because $D_n >> D_{As}$ ($\mu_n >> \mu_{As}$ and $C_{AS} >> p$ ---> $C_{As} = n$), we find (knowing that $N = n/n_i$ and by replacing n with expression [2.26] and $\vec{\nabla}$n by [2.24]):

$$\vec{E} = \frac{kT}{qC_{As}} \sqrt{\frac{\left(\frac{C_{As}}{n_i}\right)^2}{[\left(\frac{C_{As}}{n_i}\right)^2 + 4\}}} \vec{\nabla} C_{As} \qquad [2.48]$$

$$\frac{\vec{\nabla}N}{N} = \frac{\vec{\nabla}C_{As}}{\sqrt{C_{As}^2+4n_i^2}} \qquad [2.49]$$

we again find the classic expression for the internal electric field, already cited in [2.29]:

$$\vec{E} = -\frac{kT}{qC_{As}}\frac{\vec{\nabla}C_{As}}{\sqrt{C_{As}^2+4n_i^2}} = -\frac{kT}{q}\frac{\vec{\nabla}N}{N} = -\frac{kT}{q}\vec{\nabla}\,ln\frac{n}{n_i}; \qquad [2.50]$$

This is the classic expression for the internal electric field.

2.6. An example: modeling the diffusion of dopants in polysilicon grains and joints [GON 95]

Polycrystalline silicon, or polysilicon, is used, in this instance, to make MOS grids (since the early 1970s) and emitter contacts, especially in the context of BiCMOS (Bipolar Compatible CMOS) technologies. As a result of high doping using large atoms (e.g. As), the implantation may amorphize the surface, to a certain depth, of the polysilicon. During annealing, the upper part may recrystallize; therefore, diffusion is dependent on the depth.

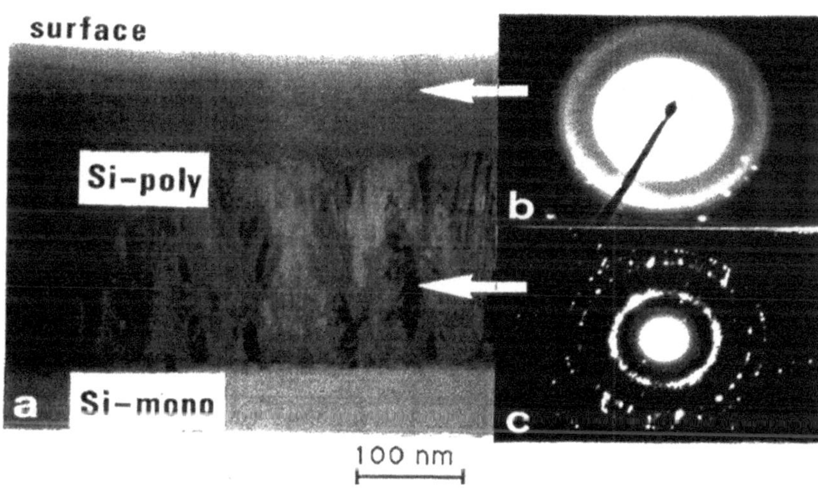

Figure 2.5. *TEM image of a transversal section of a polysilicon sample on silicon: As (10^{16} at.cm^{-2}; 100 keV) and B (10^{15} at.cm^{-2}; 30 keV) [SAD 83]*

Furthermore, fundamental electric parameters depend on the depth, typically electric conductivity. An initial approach therefore would be to divide the polysilicon layer into sub-layers; here there are two sub-layers but no restrictive number (see Chapter 3).

We will consider the diffusion in the joints and grains, taking into account the texture of polysilicon as well as the inter-diffusion between the joint/grain regions.

Let $C_{total}=C^j+C^g$

where C^j is the concentration in the joints and C^g is the concentration in the grains. We can write the following system of two equations for each dopant:

$$\frac{\partial C^g}{\partial t} = \frac{\partial}{\partial x} D^g \frac{\partial C^g}{\partial x} - T \qquad [2.51]$$

$$\frac{\partial C^j}{\partial t} = \frac{\partial}{\partial x} D^j \frac{\partial C^j}{\partial x} + T \qquad [2.52]$$

T represents the term for the transfer of dopants, from the grain towards the joint and vice versa.

These equations can be applied to arsenic and boron; thus, we can have four coupled equations:

k: trapping cste

Reaction: $X^g + S^j_{fre}$ ---------------→

←---------------

k': freeing cste

S^j_{free}: free sites in the joint: S^j_f.

X=As or B

with:

$$k = 4\pi r D^g \ [cm^3.s^{-1}] \qquad [2.53]$$

where r is the capture radius (radius of the interaction area between the free site and the dopant X; we take r = 5 Ä if there is no Coulombic interaction):

$$k' = k^*.\exp(\varepsilon/kT) \ [s^{-1}] \qquad [2.54]$$

where ε is the binding energy of the dopant–joint free site “pair”:

Trapping (t): $X^j_t = k[S^j_f - X^j].\ X^g$ [2.55]

If the sites S^j_{free} (or S{}) were considered inexhaustible (which we do not assume based on our experiments), $S_1 - X_l$ would be replaced by Si. If the two dopants, arsenic and boron, can saturate the S^j_l sites (our present hypothesis), $S^j_l - X^j$ will become S^j_l - ΣXj, with the term Σ representing a summation over all the dopants (here, arsenic and boron). This is what we will do for now in the following numerical applications:

Freeing (f): $X^g_f = k' X^j$ [2.56]

The transfer term (grains <---> joints) can be written as:

$T = X^j_t - X^g_f = k\,[S^j_f - X^j].\ X^g - k' X^j$ [2.57]

REMARK.– If the joints are columnar, we can write: $S^j_f = N_s/a$.

(Where Ns is the density of the surface free sites and a is the length of the joint.)

Let us reorganize the transfer term T:

$T = k\, S^j_f X^g - k X j X^g - k' X^j$

$= k\,[(S^j_f X^g - (k' + k\, X^g 1 X^j$ [2.58]

Thus, T can be written as:

$T = h(mX^g - X^j)$ [2.59]

with:

$h = k' + kX^g\ (s^{-1})$ (transport coefficient) [2.60]

$m = \frac{kS^j_l}{k' + kX^g}$ (segregation coefficient) [2.61]

The system can be rewritten in a more general way:

$$\frac{\partial C^g}{\partial t} = \frac{\partial}{\partial x} D^g \frac{\partial C^g}{\partial x} - h\,(m(C^g - C^j) \quad [2.62]$$

$$\frac{\partial C^j}{\partial t} = \frac{\partial}{\partial x} D^j \frac{\partial C^j}{\partial x} + h\,(m(C^g - C^j) \quad [2.63]$$

Finally, the system to be resolved effectively is written as:

$$\frac{\partial C^g_{As}}{\partial t} = \vec{\nabla}\{D^g_{As}\frac{1+ß_{As}N^g}{1+ß_{As}}\left[\left(1+\frac{C^g_{As}}{\sqrt{(C^g_{As}-C^g_B)+4n_i^2}}\right)\vec{\nabla}C^g_{As} - \frac{C^g_{As}}{\sqrt{(C^g_{As}-C^g_B)+4n_i^2}}\vec{\nabla}C^g_B\right]\} - k\left[S^j_f - C^j_{As} - C^j_B\right]C^g_{As} + k'C^j_{As} \quad [2.64]$$

$$\frac{\partial C^j_{As}}{\partial t} = \vec{\nabla}\{D^j_{As}\left[\left(1+\frac{C^j_{As}}{\sqrt{(C^j_{As}-C^j_B)+4n_i^2}}\right)\vec{\nabla}C^j_{As} - \frac{C^j_{As}}{\sqrt{(C^j_{As}-C^{gj}_B)+4n_i^2}}\vec{\nabla}C^{gj}_B\right]\} + k\left[S^j_f - C^j_{As} - C^j_B\right]C^g_{As} - k'C^j_{As} \quad [2.65]$$

$$\frac{\partial C^g_B}{\partial t} = \vec{\nabla}\{D^g_B\frac{1+\frac{ß_B}{N^g}}{1+ß_B}\left[\left(1+\frac{C^g_B}{\sqrt{(C^g_{As}-C^g_B)+4n_i^2}}\right)\vec{\nabla}C^g_B - \frac{C^g_B}{\sqrt{(C^g_{As}-C^g_B)+4n_i^2}}\vec{\nabla}C^g_{As}\right]\} - k\left[S^j_f - C^j_{As} - C^j_B\right]C^g_B + k'C^j_B \quad [2.66]$$

$$\frac{\partial C^j_B}{\partial t} = \vec{\nabla}\{D^j_{As}\left[\left(1+\frac{C^j_B}{\sqrt{(C^j_{As}-C^j_B)+4n_i^2}}\right)\vec{\nabla}C^j_B - \frac{C^j_B}{\sqrt{(C^j_{As}-C^{gj}_B)+4n_i^2}}\vec{\nabla}\right]\} + k\left[S^j_f - C^j_{As} - C^j_B\right]C^g_B - k'C^j_B \quad [2.67]$$

where $N^g = n/n_i$.

We will not dwell too long on the numerical aspect of the solution of these inter-codiffusion equations. However, we will provide some elements for clarification, which will help implement a simple algorithm.

When we study the grain to calculate one of the two concentrations (As or B) within the grain at a given time, the transfer term can be rewritten as:

$$T = -k'C^j + k(S^j_l - C^j)C^g \quad [2.68]$$

As regards the joint:

$$T = k\, S^j_f\, C^g - (k' + k\, C^g)\, C^j \quad [2.69]$$

where C^g is the concentration in the grains and C^J is the concentration in the joints.

The reason for this reorganization is that when we calculate the concentrations (of arsenic or boron) in one of the two areas (grains or joints), the others are assumed to be frozen.

Starting from a "monodiffusion" program (see the algorithm and Matlab code at the end of this chapter), we have extended the numerical solution for four coupled equations: codiffusion of boron and arsenic in the grains and joints of polycrystalline silicon, with bilateral transfer from one to the other. We fix the newly calculated concentrations (As or B) at the instant n and calculate the others (B or As), which are simply the concentrations at the instant (n-1). This back-and-forth is done at least twice for each step in time based on a fixed convergence criterion.

When we study codiffusion in the grain, we first ensure there is convergence. We then move on to the joint. Finally, we provide the solution in an inverse Gauss matricial algorithm.

When the couplings between these equations are strong, a Newton–Raphson method seems appropriate, via Jacobian matrices, following the general algorithm, $\forall\, t$:

$$C_i^{n+1} = C_i^n + A(C_i^{n+1)} C_i^{n+1}.S(C_1^{n+1}, C_2^{n+1}) \text{ for i=1or 2}$$

– Blocks method

$$C_i^{n+1} = C_i^n + A(C_i^{n+1)} C_i^{n+1}.S(C_1^n C_2^n)$$

– Nonlinear iterations

$$C=A(C)C+S(C)$$

$$[1-A(C_{j)})](C_{j+1}) = S(C_j)$$

– Gauss–Seidel method

In what follows, we present some numerical experiments to validate, in a certain manner, the coherence of the equations developed previously. At certain typical times in the annealing, we display a snapshot report of diffusion profiles in the polysilicon.

The implanted dopants are assumed to be found initially wholly in the grains. Implantation profiles, measured using secondary ion mass spectrometry (SIMS), help validate the model. The intra-grain diffusivity is that of monocrystalline silicon. Diffusivity in the joints is taken as the diffusivity in the region that is richest in the joints, that is, close to the interface.

In addition, we have taken *ab initio* the density of the free sites S to be equal to $5 \times 10^{19}.\text{cm}^{-3}$, and a freeing constant $k' = 10^{-2}\exp(-0.1/kT)$. The thermal budget is 1,050 +C - 20 s.

– First step: time: 0, 1, 2 (Figure 2.6):

The annealing starts with a rise in temperature (see the score on the temperature profile in Figure 2.6). Arsenic goes into the joints very quickly and much faster than boron. The profiles have a Gaussian progression. On the contrary, the profiles in the grains correspond to the initial implantation. The rise in temperature (the ramp) thus has an influence, which would not be the case for monocrystalline silicon alone.

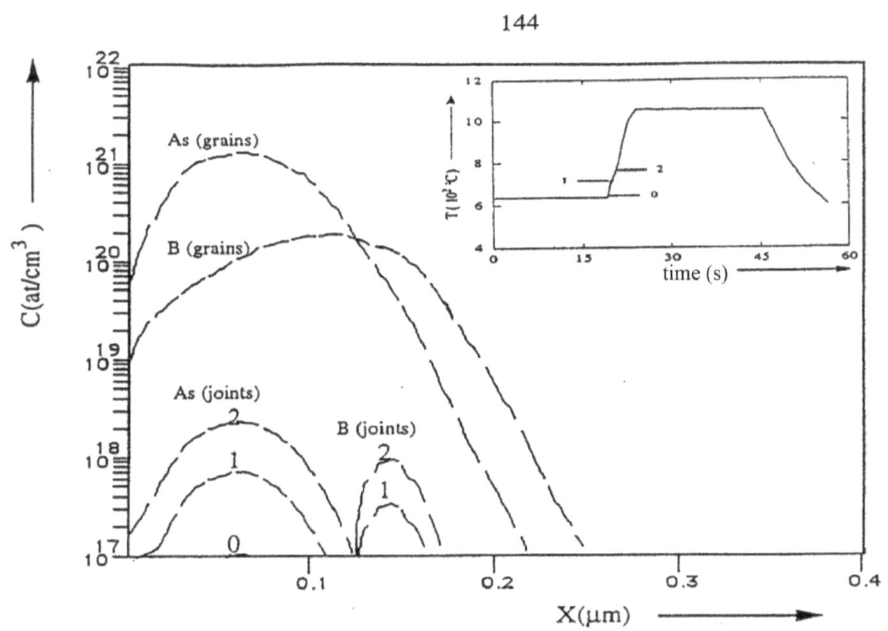

Figure 2.6. *Start of the rise in temperature*

- In the middle of the temperature rise, the plateau begins to appear.

If the concentration of boron appears at the level of the joints, it is poorly diffused in the grain. We can see an electric field effect in the grains; at the level of the intersection of the two profiles, this field retains boron at a depth of about 0.12 µm, hence the dip in the profile for boron.

In the case of monodiffusion, for arsenic, for instance, electrons (more mobile than their positively ionized donor atoms) tend to proceed. The electric field that is created will thus accelerate the arsenic and slow the electrons. The intensity of this field adjusts itself such that the macroscopic electron current is null. However, this field will at the same time slow down the negatively ionized acceptor atoms.

This phenomenon of retention between two types of dopants, such as arsenic and boron, can be found in monocrystalline silicon, as mentioned previously.

As regards the form of the profiles in the grains, it is always close to that of ion implantation.

– Second step: annealing plateau begins (here at 1,050°C):

- In the joints:

On the surface, there is a flat profile for arsenic, strong diffusion for boron and the profile is a bell curve.

- In the grains:

The profiles have not moved much in the upper part of the polysilicon (from 0 to 0.2 pm), except perhaps at the level where they cross. This is because arsenic slows down boron.

However, in the underlying part, there is a notable phenomenon. This is the freeing of arsenic at the joints; the arsenic re-penetrates into the grains, which results in a strong distribution tail in its profile. Indeed, if we interpret the transfer term, T, the codiffusion equations [2.51] and [2.52], arsenic penetrates into the joints all the more when there is a large concentration in the grain, that is, at the beginning of a new implantation peak, situated in the upper part of the polysilicon. Arsenic thus diffuses strongly in the joints, then empties into the grain wherever the concentration is low (the term in k' for T tends to compensate the term in k, and perhaps even exceeds this term; it tends to diminish as the free sites in the joints are fewer in number).

– Third step: the middle of the annealing plateau (Figure 2.7):

- in the joints: arsenic is nearly flat, as is boron, although to a lesser extent; the ratio of concentrations, in favor of arsenic, is in the order of 10;

- in the grains: arsenic presents profiles that are almost flat on the surface (= 5.10^{20} at.cm^{-3}) and in depth (= 2.10^{19} at.cm^{-3}).

A finer analysis over time would show us that there was quite a lot of back-and-forth between the grain and the joint

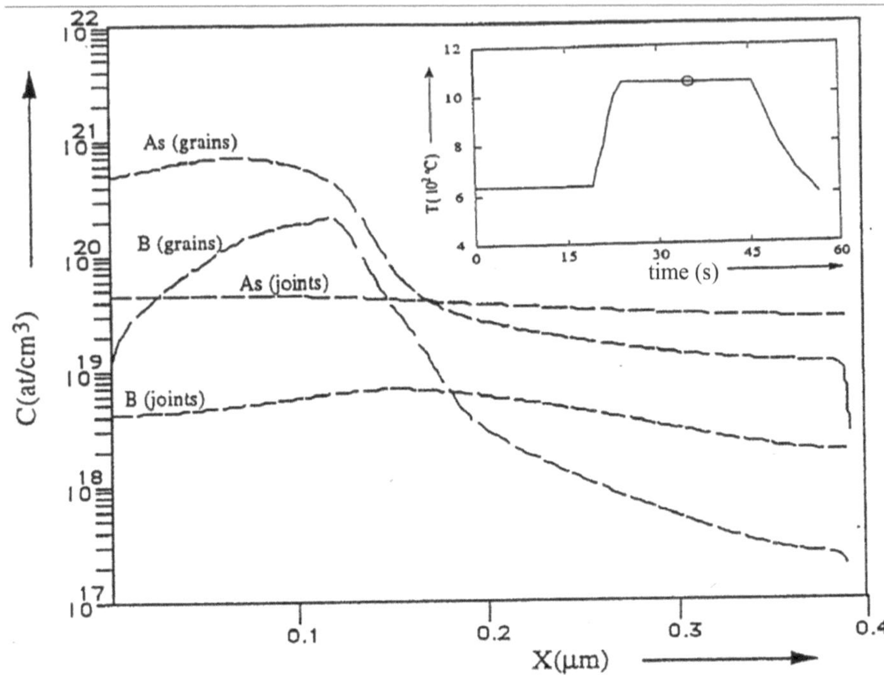

Figure 2.7. *Middle of the plateau in temperature*

For boron, still in the grain, we observe the same phenomenon as for arsenic (in the previous step), but delayed. In particular, boron atoms are freed, hence the distribution queue.

– Fourth step: end of the plateau, middle of the steeping time and, *in fine*, end of the annealing (Figure 2.8).

The phenomena are quite linear from the third step until the end of the annealing. The profiles in the joints flatten out and the freeing continues as does the diffusion. We also see that the drop in temperature is propitious for the unwinding of joints (like unscrewing a bottle of wine (or whiskey)) in the deeper part of the polysilicon.

If we compare the profile calculated at the end of the annealing of the SIMS, we can immediately see that the arsenic profile calculated in the grains is far from being more or less flat throughout, as in reality.

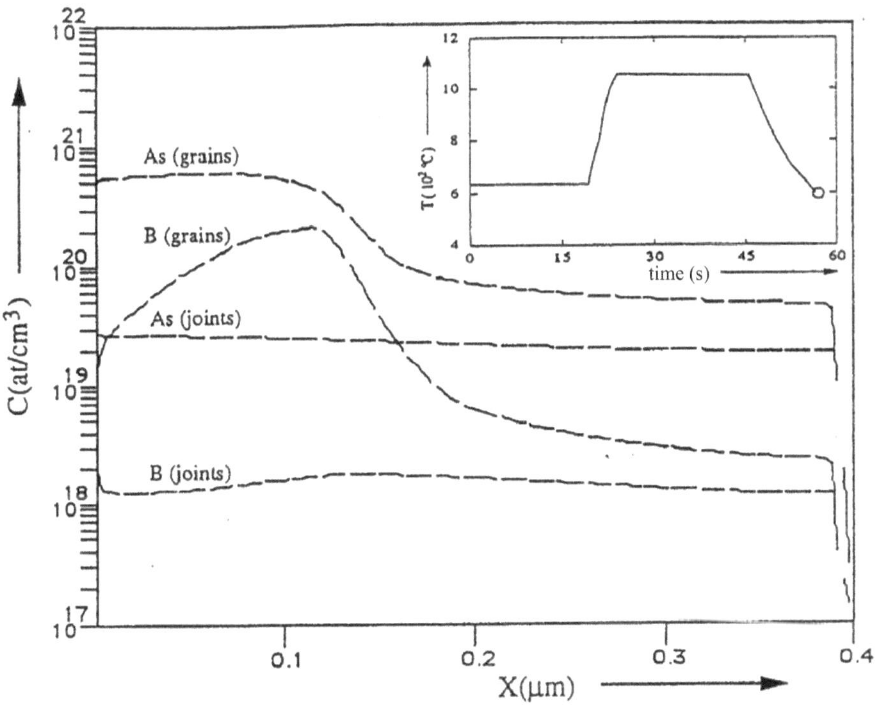

Figure 2.8. *End of the annealing*

Taking into account these remarks, let us fix $Sj = 2 \times 10^{20}\,cm^{-3}$ and $k' = 10^{-1}$ exp(-O'/kT)'. This time, there is agreement in both directions. Incidentally, we also see that the final concentration of arsenic in the joints is low (=3.5×10^{17} at.cm^{-3}), the boron concentration does not appear and is thus lower than 10^{17} at.cm^{-3}. It is important to emphasize that the "test" aspect of these calculations is in line with the fact that boron could not be detected in the joints by many different authors after annealing (for monodiffusion, in any case). The final arsenic profile in polysilicon is thus well captured. As for the boron profile, if the concentration level at the interface is obtained experimentally and by calculation (=5×10^{18} at.cm^{-3}), then it appears that the simulation describes a weaker diffusion beyond the residual hump of the implantation peak (on the surface). We can return to the Si and k' parameters.

For example, to limit the surface diffusion of boron, we reduce the density of the free sites of the joints; this idea is coherent enough if we consider that the grains are larger due to the rapid recrystallization of this zone on the surface, pre-amorphized by the arsenic implantation.

We fix:

– $Sj = 1.5 \times 10^{20}$ cm^{-3}, in the pre-amorphized zone;

– $Sj = 2.5 \times 10^{20}$ cm^{-3}, in the depth of the crystal.

In this case, we can see a deceleration in the diffusion of boron on the surface. At the interface, boron no longer attains the SIMS value. What we can see is the "progress" of profiles, around a depth of 0.18 mm, which corresponds to the abrupt variation in S_i. This drastic stress on S_i is not very physical; a linear variation of S_i (see the inset in Figure 2.9), depending on the depth, softens this transition for boron.

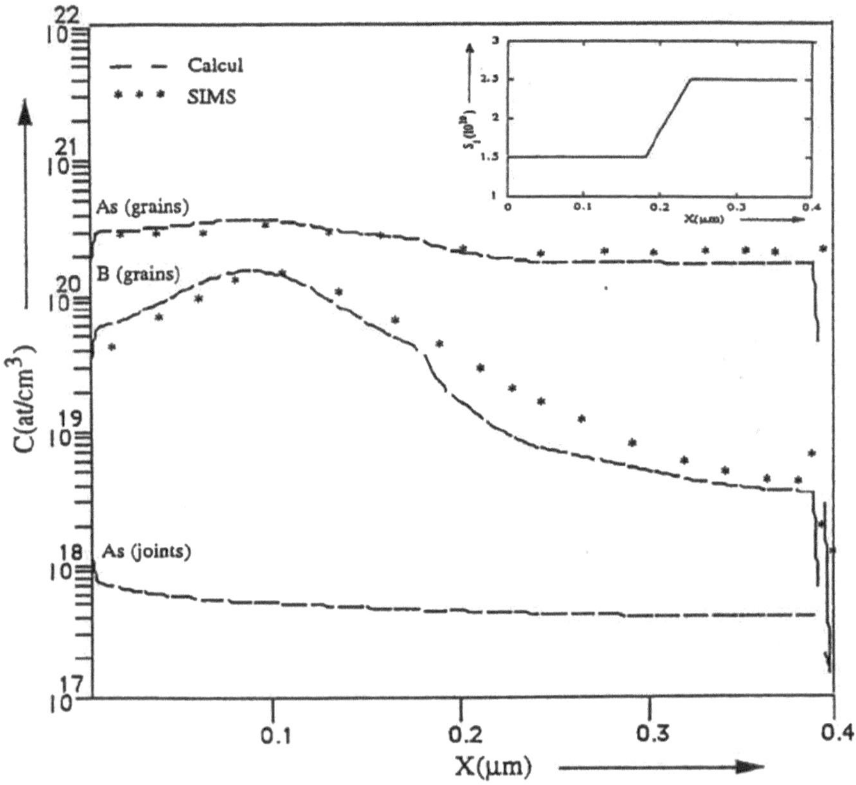

Figure 2.9. *End of the annealing, with the joints concentration not constant in depth*

This type of modeling, dividing the polysilicon into strata, or the substrate of chips, may also be adapted to the electric and thermal behavior of the silicon or the chips (see Chapter 3).

We can introduce more physics into the dynamic description of the polysilicon and its texturing [MAN 81]:

– either by describing the variation in the size of the grains over time, in an analytical manner;

– or with an increase that takes into account the dynamics and kinematics of the joints: explicitness of the forces resulting in the displacement of these joints, and taking into account this evolution. When the concentration in the grain exceeds the solubility limit (if this is known), the transport of the joint towards the grain must be canceled;

– we introduce "more physics" [MAN 81]. These authors describe the phenomena of segregation at the interfaces or surfaces based on Gibbs' adsorption equations.

If we now wish to consider the infra(micronic) dimensions, meso or nano, we must examine:

– the microscopic diffusion processes for atoms, even the defects in a broad sense (metallic defects, etc.), in a silicon-type matrix (or matrices of other kinds), at the theoretical level, irreversible processes, typically enthalpy and entropy functions;

– particular methods, statistics, abandoning the usual continuum equations such as the diffusion equation, bringing in random events.

2.7. Microscopic processes

In the case of chemical diffusion, most often, there is the proportionality between the flux of particles and its gradient. More generally, any system moves towards its minimal free enthalpy (or Gibbs energy) state; this is the chemical or even the electrochemical potential: the Fermi level for semiconductors, which controls diffusion, serving as a gauge.

What happens to the flux if the medium is subject to a temperature gradient? We then enter the realm of thermodynamics of irreversible processes.

Other than the direct effects of the heat flux and electric current, we have cross-phenomena. Hence, the necessity for writing Fick's laws in all their generality; Fick's laws, Ohm's laws and Fourier's laws are only specific cases. It is the thermodynamics of *irreversible* processes that links heat flux, material flow and electricity flow to the forces that cause them: electric, heat, stress or strains, with the last two being very strong at the microscopic level as they are of the order 1/r.

In an isolated system (which only has heat exchange with its exterior), entropy is conservative if the processes are reversible.

This is no longer the case for irreversible processes.

The conservation equations (for material, heat, etc.) result in equations of the form:

Js = JX J: flux, X: forces (the Onsager relation)

Can we easily return to Fick's law?

– Random movements:

- a particle's path does not depend on the paths of other particles;
- the jumps the particles make are independent of each other.

2.7.1. *Brownian motion [MAS 08]*

This has been well known since the theoretical demonstration by Albert Einstein (Annus Mirabilis: 1905) and Marian von Smoluchowski (1906), and since Jean Perrin's experiments.

Brownian motion was discovered in 1827 by the botanist Robert Brown. He observed under the microscope that pollen grains dispersed in water were subject to a continual and irregular movement.

The properties of this random walk result in the particle remaining centered, on average, on its starting point (with the mean value of displacement being null); however, with time, the particle may "explore" points that are farther and farther away (Figure 2.9); the square of displacement (variance) increases linearly with time. When the particle is larger than a few micrometers, Brownian motion is no longer observed: this is because there are too many atoms whose speed is randomly distributed and bombards the target. Brownian motion is therefore one of the characteristics of colloidal suspensions. It is the fundamental unit needed to construct the continuous Markov processes (over R or Rn) that we call diffusions. It makes it possible to construct stationary Gaussian processes.

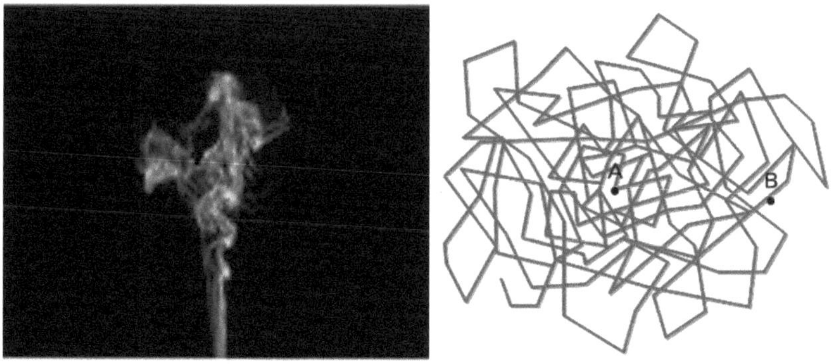

a) Turbulent smoke b) Brownian motion

Figure 2.10. *An example (a) and a schematic trajectory (b) of Brownian motion. (http://www.universalis-edu.com/encyclopedie/mouvement-brownien/). For a color version of this figure, see www.iste.co.uk/gontrand/nanoelectronics.zip*

For one dimension, let X be a path with a mean null value, $\langle X\rangle = 0$, and the mean square value deviation $\langle X^2\rangle = \sigma^2$. The particles in the form of a Dirac distribution at x= 0 and t= 0 diffuse from left to right.

For any T < t, there is a corresponding path x or – x.

Let C (respectively C'_1) be the concentration at + x (respectively - x).

The flux can then be split in two parts: ½.C. x (flux to the right) and ½.C'. x (flux to the left); the net flux of the particle is therefore:

$$\tfrac{1}{2}.(C - C')x \qquad [2.70]$$

The mean concentration between 0 and +/– x1 is ([C - C')/x, an expression that can be likened to the derivative:

$$\frac{dC}{dx} = \frac{C'_1 - C_1}{x} \qquad [2.71]$$

with:

$$D_{x,y,z} = \frac{\langle x^2\rangle}{2t} \qquad [2.72]$$

The dimension of $D_{x;y;z}$ represents a surface per unit time (spreading).

The number of particles diffusing per unit time between −x and +x, per unit area of the planes passing through −x and +x, planes perpendicular to the line passing through these two points, is:

$$J_{x,y,z} = \frac{1}{2t}(C - C')x = \frac{1}{2t}x^{2}\frac{.dC}{dx} \qquad [2.73]$$

This equation is the equation relative to Fick's law, with diffusivity being:

$$D_{x,y,z} = = \frac{1}{2t}x2 = \frac{1}{2t}<X2> \qquad [2.74]$$

For a cubic structure, the dopant may skip any of the three directions, with the length of the jumps being equal because of the symmetry. Hence:

$$< L^2 > = 3 < X^2 > \qquad [2.75]$$

Therefore, the diffusion coefficient is isotropic:

$$D = \frac{<L^2\,(t)>}{6t} \qquad [2.76]$$

In silicon (e.g. face-centered cubic; FCC), again, because of the symmetry, the length of the jumps does not depend on the direction, and we can write:

$$< L^2(t) \geq n\delta l \qquad [2.77]$$

with n being the number of successive jumps and δl being the length of the elementary jump.

Where n = F. t, with F being the frequency (number of jumps) per unit time.

In the case of silicon (e.g. FCC):

$$\delta l = a.\sqrt{2}/2 \qquad [2.78]$$

where a is the size of the crystalline mesh.

Finally, we thus have:

$$D = F\ a2 \qquad [2.79]$$

2.7.2. *Diffusion mechanisms and correlation effects*

The hypothesis of random motion (independent of successive jumps of the diffusing particle) is not always verified – this depends on the nature of the particles and the mechanism of the jumps; there are, however, correlations between the jumps.

How does an atom migrate within the lattice? Figure 2.11 gives a recap of the simplest mechanisms that allow this migration.

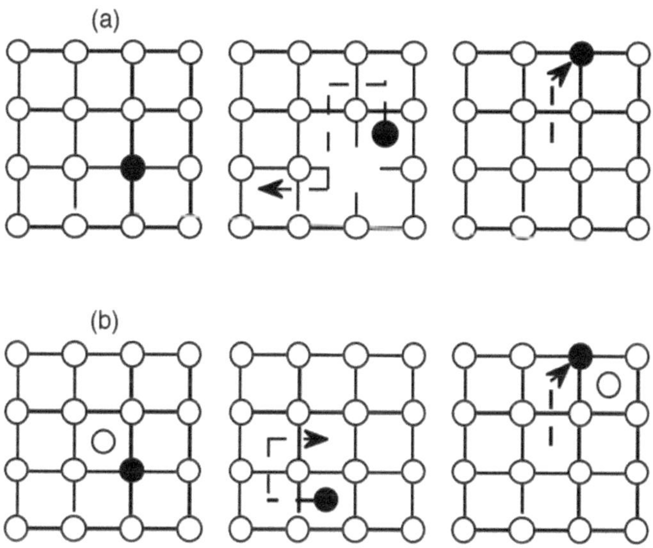

Figure 2.11. *Dissociative mechanisms [MAT 96]: (a) "Frank–Turnbull"; (b) "kick out"*

2.7.2.1. *Interatomic exchange*

For an atom in a lattice site, there may be a simple exchange mechanism between two atoms or a cyclic exchange mechanism that brings into play three atoms or even more. The first is not very probable, given the strong repulsion between atoms over a short distance, which forbids the intermediary position, with both atoms thus being mid-way. In the cyclic exchange, on the contrary, the repulsive forces play an active role – each atom pushing its neighbors during a circular type of permutation. It requires more coordinated atomic jumps, which makes it less probable. If it allows a migration of a tracer atom, it cannot take into account a transport of material in one direction, as is observed in chemical diffusion, creep, sintering and so on. For the diffusion of the tracer in ordered alloys, certain cycles that preserve the order are brought in, unlike mechanisms that are used occasionally.

2.7.2.2. *Assistance through occasional defects*

Most diffusion mechanisms require the presence of occasional defects.

2.7.2.2.1. Vacancy mechanism

If a site is not occupied, a nearby, neighboring atom may jump to this site (note: a perfect crystal cannot thus be ideal), thereby causing the appearance of a vacancy at the site from which it left; the number of vacancies is conservative: we can have migration of the vacancy and/or migration of the atom. In a crystal formed by a pure element, at any moment, the vacancy may jump towards any of the first neighboring sites. If these successive jumps are independent of each other, it is not the same for a marked atom – either an isotope of the atoms of the crystal (self-diffusion) or an atom that is chemically different (heterodiffusion; see dopants).

2.7.2.2.2. Self-diffusion

Let there be an initial jump by the identified atom, situated at A towards the vacancy site B. After this jump, the atom situated in B finds itself adjacent to the vacancy in A; a second BA jump is possible, with a higher probability than the jump that would result from a random distribution. In calculating the path, we have two jumps without a net displacement. Other eventualities are possible; the vacancy positioned on A may jump towards any of the S first adjacent sites.

The probability that the second jump of the marked atom will cancel the first is 1/S. Over n jumps, there will be n(1/S) pairs of this type, that is, 2.n(1/S) failed jumps or n.(1-2/S) effective jumps; that is, those that lead to random migration. We will consider an effective jump frequency $\Gamma eff = f\Gamma$, with $f = 1\text{-}2/S$, a quadratic path and a diffusion coefficient reduced by the same factor with respect to the values calculated for the random case. F is called the correlation factor, which is between 0 and 1.

2.7.2.3. *Diffusion mechanisms and correlation effects*

Between two successive jumps for a vacancy mechanism:

The earlier description is highly simplified, as the vacancy situated at A is able to migrate to several successive sites and, after jumps, able to return to one of the sites adjacent to the considered atom and thus exchange with that atom; the second jump the atom makes is not independent of the first.

2.7.2.3.1. *Interstitial mechanism*

An interstitial (or anti-site) atom jumping from one interstitial site to another is an occasional defect whose successive jumps are not correlated. This is called the direct interstitial mechanism. It is typical of atoms that are normally found in the solid interstitial mechanism. A variant of this mechanism exists if the marked atom can find itself in a substitutional and interstitial position – this is called the indirect interstitial mechanism.

If the atom is in substitution (on site), it can jump only if an interstitial arrives in a neighboring position and chases it away from its site, thereby taking up the interstitial position itself. This atom may then jump towards any adjacent substitutional site, thus creating an interstitial defect. The configuration at this stage resembles the case of the vacancy mechanism as an impurity (an interstitial) is adjacent.

The reverse jump of the interstitial resets the configuration to the previous stage: the marked atom in the interstitial position; these two jumps are correlated.

However, the correlation effect stops here because as soon as the considered atom is in the interstitial position, there is no longer (within the configuration) any memory of the previous jump (see Markovian process) and all possible jumps are equiprobable. The correlation effects are limited to pairs of jumps: insertion–substitution–insertion, while in the vacancy mechanism, all successive jumps are increasingly closely related. The situation is a little more complicated in reality, as two jumps by two atoms brought into play in the elementary process are not always collinear: there are many types of jumps to be considered depending on the arrangement of the interstitial sites around a substitutional site. Another complication arises from the fact that the configuration of the defect may be more complex: this is the dissociated interstitial, which appears in the form of a pair of atoms centered on a site (see dumbbell). Three jumps are needed to go from one site to the adjacent sites: the first incorporates the atom into a dumbbell, the second corresponds to the jump of this atom and, in the third jump, the atom returns to the substitution site.

2.7.2.3.2. *Bivacancy mechanism*

The atom is interchanged with the vacancies of the pair. In the FCC structure, this is possible without dissociating the pair of vacancies. However, in the FCC structure, the jump necessarily results in a modification of the distance between the two vacancies, as we cannot find three sites that are all in the position of the first adjacent sites in this structure.

2.7.2.3.3. *Other mechanisms*

There are also more complex mechanisms. The occasional defect may be less localized than in the classic representation. We thus arrive at a concept of an extended interstitial, which, if it involves a dense row of atoms, is called a "crowdion" or extended vacancy that, up to a limit, becomes a small, "relaxed" zone of disorder.

This defect has been described as a local fusion – the atoms are able to migrate by a "rushing" movement, comparable to the motion of atoms in a liquid. Complexes of occasional defects have been proposed in specific cases: interstitial–vacancy pair, with the interstitial being in a dissociated form and so on.

REMARKS.–

Several elementary mechanisms may compete: the considered atom being able to jump thanks to one or more of the mechanisms previously described.

The correlation effect between two successive jumps appears as soon as at least three particles enter the process.

– *Diffusion mechanisms and correlation effects*

For this purpose, atoms in the lattice must be considered to be particles whose natures differ – this includes the marked atoms, the vacancies, the interstitials and the self-interstitials.

– *Definition for the correlation factor*

The theory of random movement leads to mean quadratic path values $<X^2>$. In the presence of the correlation effect, the means of the products $<X_iX_j'>$ are no longer null. We thus write, by definition:

- For a purely random movement: $<X_iX_j> = 0$:

- If the successive jumps of an atom are caused by two different defects, they are not correlated. The $<X_iX_j>$ are then null if the jumps i and j are due to two different defects.

2.7.2.4. *Application to cubic structures (for example, Si)*

In a direct interstitial mechanism, only two consecutive jumps may be correlated.

2.7.2.4.1. The calculation algorithm

In order for the vacancy concentration to be constant, each time that it emerges from the crystal, we reintroduce it through the opposite face (period conditions, which are the same as considering the crystal to be of infinite dimensions).

The random movement of the vacancy is followed by a random draw (see Monte Carlo-type simulation) for a number for each jump; we use a procedure that uniformly yields (pseudo)-random numbers between 0 and 1. For the random draw, the segment [0,1] is divided into six equal parts, corresponding to the six possible directions for the jump.

The number picked fixes the direction of the jump. We note the new coordinates of the vacancy and the concerned atom. We then recommence the operation. *In fine*, the vacancy has carried out N jumps. Each atom has been numbered, and it is possible to know, for each of the number of jumps, m^+ and m^-, that it took place in the directions x and −x. We thus obtain the distribution W(m, n), taking into account the number of atoms that have followed a path $m\lambda = (m^+ - m^-)$.

The result is a Gaussian. We can also verify the correlation effects by calculating $<X^2>$ and using the quadratic path on a large number of atoms (several thousand).

It is thus easy in such a program to introduce the effect of a transport force.

We must then carry out two random draws for each vacancy jump (random numbers uniformly distributed over the segment [0,1]). The first number, as mentioned, indicates the direction of the jump. The second indicates whether the jump has really taken place or not, $0 < r < 1$; if the number drawn is larger than $R = \exp(-W'/kT)$, the jump has not taken place. If it is smaller than or equal to R, the jump has taken place, and we recalculate the coordinates of the vacancy and the atom with which it has changed places.

We can verify the Nernst–Einstein equation (v=FD/kT; F being the transport force and D being the diffusivity). To do this, we must compare $<x>$ with $<x^2>$, this variance being evaluated in the absence of any field if we take $<x^2>$; in the presence of a field, we use $<x^2>_{E=0} = <x^2>_{E\#0} - <X>^2$.

2.7.2.4.2. Kinetic method

This method is called the analogy related to the kinetic theories of transport in fluids. For the diffusion of the impurity, we directly carry out the calculation for the flux of atoms in a stationary concentration gradient based on kinetic equations for different impurity–defect pairs. Hence, one expression: flux based on different frequencies and jump lengths, and the value of f by identification with Fick's law.

2.7.2.4.3. Analog simulation method

This method is based on the analogy between Fick's laws and Ohm's laws: electric currents may simulate diffusion flux using an electrical network of equivalent resistances. In certain cases, it is possible to solve equations; in more complex cases, we construct the equivalent network and carry out the appropriate electric measures. This method sadly seems obsolete.

2.7.2.4.4. Numerical simulation method

This method has been extensively developed since the 1940s–50s thanks to increasingly powerful computers. It uses a random draw (see particular methods, Monte Carlo-type method) to simulate atomic trajectories and $<X^2>$.

NOTE.– The correlation fact is a purely geometric numerical factor, determined by the structure alone, and thus independent of temperature – but only when there is a single jump frequency for the defect. This can only be realized for self-diffusion (identical defect-atom exchange frequency, whether it is an atom in the lattice or an isotope, up to the isotopic effect).

NOTE.– In non-cubic structures, f, like D, is a tensor. It comprises 3 r components: 3 for the three axes and r for the number of jump types.

2.7.2.4.5. Interactions between defects or between diffusant particles

The jump frequencies increase or reduce locally depending on the environment of the defect. A simple case is that of neighboring sites excluded following repulsion between defects. In a barrier-potential model, minimum energies are not always at the same level, resulting in an additional term in jump frequency, for example:

$$w = \exp(-\Delta H/kT)\ \text{exP}(\xi\Phi/2kT) \qquad [2.80]$$

where ξ is the number of nearby sides adjacent to the site where the particle may jump, sites where there are identical particles with which there is a Φ interaction (limited to the first neighbors, for simplification).

In reality, we cannot isolate the particle that jumps from the system formed of neighboring atoms; this is therefore an n-body problem, which is very difficult to approach. Two equivalent approaches have been proposed:

– developing an n-body calculation based on statistic mechanics of equilibrium. The peak of the potential barrier defines a certain state of the system, with a given configuration and modes unique to vibration, different from that of the stable state;

– a "dynamic" approach based on stochastic theories, calculating the probability of the superposition of phonons such that the migrating particle and the neighboring atoms simultaneously present displacements with amplitudes and directions such that the jump is allowed.

2.8. Phenomenological theory of diffusion

Ab initio, the fundamental relation for diffusion ("Fick") was introduced as a law, emphasizing its formal analogy with other laws of physics – and transport terms have been added without justification to the nature of the transport force. Even though atomic models based on random movements of the diffusant species have confirmed these laws, it is interesting to study the fundamentals in a more rigorous manner in order to precisely write the equations corresponding to the experimentation, depending on the number and nature of the constituents, the types of forces and so on. As diffusion is essentially an irreversible process, the theory of irreversible processes (TIP) comes into play [ADD 66].

2.9. Review of the thermodynamics of irreversible processes

The physical quantity that characterizes processes when there is irreversibility, whether the system is open or closed, is entropy:

$$dS = \frac{dQ}{T} + dS_{irr} \qquad [2.81]$$

where dQ is the heat exchange with the external medium (thermostat) and t is the time:

$$\frac{dS}{dt} = \frac{dS_e}{dt} + \sigma \qquad [2.82]$$

where $\sigma = dS_{irr}$ is the source of the entropy and dS_e is the conservative part. If there is local equilibrium, we calculate σ by applying Gibb's relation to a small volume of material, and write the conservation laws: $\frac{\partial C_i}{\partial t} = -divJ_i$, for different constituents and for energy. Entropy is thus divided into non-conservative and conservative parts:

$$\frac{dS}{dt} = -divJ_s + \sigma \qquad [2.83]$$

The main result of this calculation is that the entropy source, σ, can be written as:

$$\sigma = J_q.F_q + \Sigma_k J_k.F_k \qquad [2.84]$$

that is, the sum of the products of the flux J with the corresponding forces.

Where:

J_q is the heat flux;

J_k is the flux of the constituent k;

$X_q = \vec{V}(1.T)$.

$$X_k = \frac{F_k}{T} - \vec{\nabla}(\mu_k / T) \qquad [2.85]$$

F_k is the external force applied to the constituent k (electric field, for example). We then write:

$$X_k = -\vec{\nabla}(\widetilde{\mu_k})/\mathrm{T}) \qquad [2.86]$$

with the generalized chemical potential $\widetilde{\mu_k}$. If Fk derives from an electric potential ϕ, then $\widetilde{\mu_k} = \mu_k + q_k$. Here, $\emptyset$ is the electrochemical "potential" and q_k is the electric charge.

When there is no force X: $\sigma = 0$, there is no irreversible process; thus, the J fluxes are null. Depending on the forces, the fluxes are written so as to respect this condition ($\sigma = 0$):

$$J_i = \sum_{k=1}^{p} \quad L_{ik}.X_k + L_{iq}.X_q \qquad [2.87]$$

$$J_q = \sum_{k=1}^{p} \quad X_{qk}.X_k + L_{qq}.X_q \qquad [2.88]$$

The L coefficients, called the phenomenological coefficients, are independent of the forces. The diagonal coefficients may be atomic mobility, thermal conductivity and so on. The other coefficients are related to the coupling phenomena (thermos transport or the Soret effect): processes out of equilibrium, which occur when a dense fluid mixture is subject to a temperature gradient. The response to the system is the appearance of molar fraction gradients, the Thomson effect (a homogeneous conductor through which a current travels provokes a heat flux – or a temperature gradient) and the Dufour effect (symmetric to the Soret effect), i.e. heat diffusion due to a concentration gradient, via the principle of micro-reversibility, and the flux and forces are calculated based on the entropy source, given by the Onsager relation:

$$L_{ik} = L_{ki} \qquad [2.89]$$

More precisely, each flux is expressed based on the forces of the same tensor order (order 2: vector) in the case of equations [2.87] and [2.88]. If the system is the site of chemical reactions, we have scalar terms: reaction speed and the corresponding oversaturation force.

In a system at mechanical equilibrium, the forces verify the Gibbs–Duhem equation:

$$\sum_{k=1}^{p} \quad \rho_k.(\vec{\nabla}\tilde{\mu}_k)_T = 0 \qquad [2.90]$$

where ρ_k is the mass concentration (per unit volume) of the species k. The entropy source is then written as:

$$\sigma = J_q . X_q + \sum_{k=i}^{p-1} J_k . (X_k - X_p) \quad [2.91]$$

Any flux expressed as a function of (p - 1) independent forces is written as:

$$J_i = \sum_{k=i}^{p-1} L_{ik}(X_k - X_p) \quad [2.92]$$

These material fluxes are defined with respect to the barycentric speed:

$$\omega = \sum_k \frac{\rho_k v_k}{\rho} \quad [2.93]$$

where ω represents the average speed and v_k represents the speed of the constituent k in the fixed referential R^0. It then follows that the fluxes are defined in the referential R' by:

$$J_k = \rho_k (v_k - \omega) \quad [2.94]$$

Hence:

$$\sum_k J_k = 0 \quad [2.95]$$

In general, the measures are put in place in the fixed referential, where the $\mathring{J_k}$ fluxes dependent on one another are written as:

$$\mathring{J_k} = \rho_k v = J_i + \rho_k \omega \quad [2.96]$$

The above formalism must be applied with some precautions. In particular, we must take into account occasional defects. These are the null mass constituents for vacancies. On the contrary, vacancies and interstitials differ fundamentally in chemical species as they may be created or destroyed. The result of these observations is that the referential in which we express the J fluxes is taken as a referential related to the mean atomic speed rather than the mean barycentric speed. The mean atomic speed is a quantity that is accessible in experiments (inert markers, for example). For a purely vacancy-based mechanism, to which we will be limiting ourselves in the following sections:

$$\mathrm{V} = \frac{n_l v_l + \sum_1^p n_k v_k}{n} \quad [2.97]$$

where v_k and v_l are respectively the mean speeds of the k atoms measured and the vacancies in the fixed referential Ro.

The entropic balance (see Gibbs equation) – the entropy of an ideal mix of perfect gases is the sum of the entropies of its constituents, assumed to be separate, at the mixture temperature and under pressures equal to the partial pressures they exert in the mixture – leads to:

$$\sigma = J_q.\vec{\nabla}\left(\frac{1}{T}\right) - \sum_{l-1}^{p} J_j.\vec{\nabla}\left(\frac{\tilde{\mu}_k-\tilde{\mu}_l}{T}\right) - \frac{R_l-\mu_l}{T} \quad [2.98]$$

From [2.98], we have the phenomenological relations for the scalar and vector quantities in a system with p material constituents; the vacancies form the $(p+1)^{th}$, and the fluxes are expressed as the number of particles per unit time and area:

$$J_k = L_{kq}.\vec{\nabla}\left(\frac{1}{T}\right) - \sum_{l-1}^{p} L_{ki}.\vec{\nabla}\left(\frac{\tilde{\mu}_k-\tilde{\mu}_l}{T}\right) \quad [2.99]$$

$$J_q = L_{qq}\vec{\nabla}\left(\frac{1}{T}\right) - \sum_{l-1}^{p} L_{ki}.\vec{\nabla}\left(\frac{\tilde{\mu}_k-\tilde{\mu}_l}{T}\right) \quad [2.100]$$

$$L_{ik} = L_{ki} \quad [2.101a]$$

$$L_{qk} = L_{qk} \quad [2.101b]$$

$$R_l = \acute{a}\mu_l/T \quad [2.102]$$

$$J_i = \sum_{k=1}^{p} J_k \quad [2.103]$$

The last equation can be deduced from the definition of the mean atomic speed.

Indeed, in the mobile referential R, the conservation equation for a conservative material species can be written in the referential R_0:

$$\frac{\partial n_k}{\partial x} - div(n_k v_k) - div J_k^0 \quad [2.104]$$

Hence:

$$\sum_k \frac{\partial n_k}{\partial x} = -div \sum_k (n_k v_k) \quad [2.105]$$

If the total number of sites per unit volume, n, verifies:

$n = n_l \sum_1^p n_k = \text{constant},$

then:

$$\frac{\partial n_k}{\partial x} - div(n_l v_l) - div(nv) = divJ_k^0 + R_l \quad [2.106]$$

where $R_l = div(nv)$ represents the number of vacancies created per unit time and volume.

1) For a system subject to an electric field, we must *a priori* take into account the constituents, like the electrons or holes, which form the $(p + 2)^{th}$ constituent:

$$J_k = L_{kq}.\vec{\nabla}\left(\frac{1}{T}\right) - \sum_{l-1}^{p} L_{ki}.\vec{\nabla}\left(\frac{\tilde{\mu}_i - \tilde{\mu}_l}{T}\right) - L_{ke}.\vec{\nabla}\left(\frac{\tilde{\mu}_e}{T}\right) \quad [2.107]$$

$$J_e = L_{qq}\vec{\nabla}\left(\frac{1}{T}\right) - \sum_{l-1}^{p} L_{ki}.\vec{\nabla}\left(\frac{\tilde{\mu}_i - \tilde{\mu}_l}{T}\right) - L_{ee}.\vec{\nabla}\left(\frac{\tilde{\mu}_e}{T}\right) \quad [2.108]$$

$$J_q = L_{qq}\vec{\nabla}\left(\frac{1}{T}\right) - \sum_{l-1}^{p} L_{ki}.\vec{\nabla}\left(\frac{\tilde{\mu}_i - \tilde{\mu}_l}{T}\right) - L_{ee}.\vec{\nabla}\left(\frac{\tilde{\mu}_e}{T}\right) \quad [2.109]$$

$$J_i = \sum_{k=1}^{p} J_k \quad [2.110]$$

2) The Onsager relations between the phenomenological coefficients may not be verified if the mean reference speed is poorly chosen.

3) The above equations assume that all the species, including the vacancies, have the same atomic volume. This restriction can be removed via the introduction of partial molar volumes.

The phenomenological equations are very powerful. However, they quickly lead to heavy expressions with many phenomenological coefficients, whose physical significance is not always clear at first sight. This is like the statistical theories and models (random movement, Monte Carlo method, kinetic method) of the explication of expressions for the coefficients Lij in terms of elementary characteristics (defect concentration, jump frequency, interactions with the solute atoms, etc.).

2.10. Conclusion

The most crucial steps in the manufacturing process of circuits, and thus of the devices of micro-nanoelectronics, are the steps involving diffusion during the annealing. We know that the diffusion equation is related to Brownian motion; this being the case, by looking at high levels of miniaturization, we can orient ourselves towards the random walk-type statistical methods. Moreover, it seems necessary to go deeper into the microscopic mechanisms of diffusion of dopants, especially related to the presence of defects, which would no longer seem to be a problem for microelectronics alone.

A more in-depth and promising project is that of following the confrontation/comparison of the governing equations linked to the modeling of the diffusion processes.

2.11. Appendix 1

Numerical modeling of the simple diffusion

We first resolve the diffusion equation numerically. To do this, we use the finite differences method.

We divide space into k segments (in the case of 1D) with a thickness Δx and carry out the diffusion over small intervals of time, Δt (Δx and Δt may vary).

Here, in order to simplify the calculations, we do not take into account the exodiffusion term.

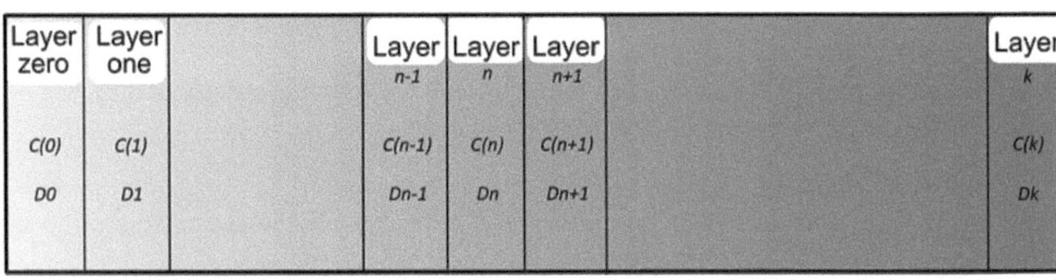

[2.111]

Thus, Fick's first law expresses the flux of dopants that passes from one segment to another:

$$F=-D\ \delta C(x,t)/\delta x \qquad [2.112]$$

Taking into account the division, the flux becomes:

$$F(k,n) = -D(k,n)*[C(k,n)-C(k,n-1)]/\Delta x \qquad [2.113]$$

Here, we thus have the flux passing from the layer n-1 to the layer n during the time interval comprised between the instants:

$$(k-1)\Delta t \text{ and } k\Delta t. \qquad [2.114]$$

Fick's second law gives us:

$$\delta C(x,t)/\delta t=\delta(D\ \delta C/\delta x)/\delta x \qquad [2.115]$$

where D depends, in particular, on C and thus on x.

This equation may be translated after the division of the medium in the following manner:

$$[C(k,n)-C(k-1,n)]/\Delta t=[F(k,n)-F(k,n+1)]/\Delta x \qquad [2.116]$$

Hence, the expression for the second law using the expression for flux from the first becomes:

$$[C(k,n)-C(k-1,n)]/\Delta t=\{-D(k,n)*[C(k,n)-C(k,n-1)]/\Delta x$$

$$+D(k,n+1)*[C(k,n+1)-C(k,n)]/\Delta x\}/\Delta x \qquad [2.117]$$

NOTE.–

We have thus obtained a relation that expresses the concentration at time k as a function of the concentrations at time n-1.

For the next part, we will posit:

$$\Delta t/(\Delta x)^2 = A_o \qquad [2.118]$$

with:

$$A(k,n) = Ao * D(k,n) \qquad [2.119]$$

hence:

$$C((k-1,n) = -A(k,n) * C(k, n-1) +[1+A(k,n) +A(n,j+1)] * C(k,n) - A(k,n+1) * C(k,n+1) \qquad [2.120]$$

Limiting conditions

We consider that the flux between layer zero (external layer) and layer one is null.

That is.

$$C(k,0)=C(k,1). \qquad [2.121]$$

Thus, in equation [2.120]:

$$C(k-1,1) = [1 + A(k,2)]*C(k,1) - A(k,2)*C(k,2) \text{ (IV)} \qquad [2.122]$$

We know from theory that the concentration in a medium tends towards 0 when x (or the variable for depth) tends to infinity.

Experimentally, we must fix this limit to a segment, k, in which the concentration becomes negligible with respect to the earlier concentration:

$$Cn+1 = Cn \qquad [2.123]$$

From [2.123], we obtain:

$$C(k-1,n)=-A(k,n)C(k,n-1)+[1+A(k,n)]C(k,n) \qquad [2.124]$$

We will now write this system of equations in matrix form, relating the concentrations to time (k-1)Δt and to those at time k Δt thanks to equations [2.120], [2.122] and [2.124].

$$\begin{bmatrix} C(n-1,1) \\ C(n-1,2) \\ \vdots \\ C(n-1,j) \\ \vdots \\ C(n-1,k) \end{bmatrix} = \begin{bmatrix} \\ \\ \textbf{MATRIX} \\ \textbf{2.126} \\ \\ \\ \end{bmatrix} \begin{bmatrix} C(n,1) \\ C(n,2) \\ \vdots \\ C(n,j) \\ \vdots \\ C(n,k) \end{bmatrix} \qquad [2.125]$$

$$\begin{bmatrix} 1+A(n,2) & -A(n,2) & 0 & 0 & 0 & 0 & 0 & 0 & 0 & 0 \\ -A(n,2) & 1+A(n,2)+A(n,3) & -A(n,3) & 0 & 0 & 0 & 0 & 0 & 0 & 0 \\ 0 & 0 & 0 & 0 & 0 & 0 & 0 & 0 & 0 & 0 \\ 0 & 0 & 0 & 0 & 0 & 0 & 0 & 0 & 0 & 0 \\ 0 & 0 & 0 & 0 & 0 & 0 & 0 & 0 & 0 & 0 \\ 0 & 0 & 0 & -A(n,j) & 1+A(n,j)+A(n,j+1) & -A(n,j+1) & 0 & 0 & 0 & 0 \\ 0 & 0 & 0 & 0 & 0 & 0 & 0 & 0 & 0 & 0 \\ 0 & 0 & 0 & 0 & 0 & 0 & 0 & 0 & 0 & 0 \\ 0 & 0 & 0 & 0 & 0 & 0 & 0 & 0 & 0 & 0 \\ 0 & 0 & 0 & 0 & 0 & 0 & 0 & 0 & -A(n,k) & 1+A(n,k) \end{bmatrix} \qquad [2.126]$$

$$\begin{bmatrix} U1 & -A(n,2) & 0 & 0 & 0 & 0 & 0 & 0 & 0 & 0 \\ 0 & U2 & -A(n,3) & 0 & 0 & 0 & 0 & 0 & 0 & 0 \\ 0 & 0 & U3 & -A(n,4) & 0 & 0 & 0 & 0 & 0 & 0 \\ 0 & 0 & 0 & U4 & 0 & 0 & 0 & 0 & 0 & 0 \\ 0 & 0 & 0 & 0 & 0 & -A(n,j) & 0 & 0 & 0 & 0 \\ 0 & 0 & 0 & 0 & 0 & Uj & -A(n,j+1) & 0 & 0 & 0 \\ 0 & 0 & 0 & 0 & 0 & 0 & Uj+1 & 0 & 0 & 0 \\ 0 & 0 & 0 & 0 & 0 & 0 & 0 & 0 & 0 & 0 \\ 0 & 0 & 0 & 0 & 0 & 0 & 0 & 0 & 0 & 0 \\ 0 & 0 & 0 & 0 & 0 & 0 & 0 & 0 & 0 & Uk \end{bmatrix} \qquad [2.127]$$

With

$$\begin{aligned} U_1 &= 1 + A(n,2) \\ U_2 &= \left[1 + A(n,2) + A(n,3)\right] - \left[A(n,2)\right]^2 / U_1 \\ &\vdots \\ U_j &= \left[1 + A(n,k)\right] - \left[A(n,k)\right]^2 / U_{j-1} \\ &\vdots \\ U_k &= \left[1 + A(n,k)\right] - \left[A(n,k)\right]^2 / U_{k-1} \end{aligned} \qquad [2.128]$$

To simplify the next section, we will first triangulate this matrix. Then, in order to explain the method used, we will first review and then explain the result.

After calculations, the following matrix is obtained:

with

$U1 = 1 + A(k,2)$

$U2 = [\,1 + A(k,2) + A(k,3)\,] - [A(k,2)]^2 / U_1$

...

$U_j = [\,1 + A(k,n)] - [A(k,n)]^2 / U_{j-1}$

...

$U_k = [1 + A(k,n)] - [A(k,n)]^2 / U_{n-1}$ [2.129]

To obtain this matrix, we have thus posited:

U1=1+A(k,2)

We then multiplied the "j-1th" line by:

A(k,j)/Uj

which we then added to the "j-[th]" line.

The factor is judiciously chosen so as to cancel the term placed under the main diagonal (we keep the first line unchanged), and we thus have the new matrix.

Consequently, this change to the matrix leads us to carry out a change in the variable (to keep our equality) such that Ck-1 becomes Kn-1.

We then obtain:

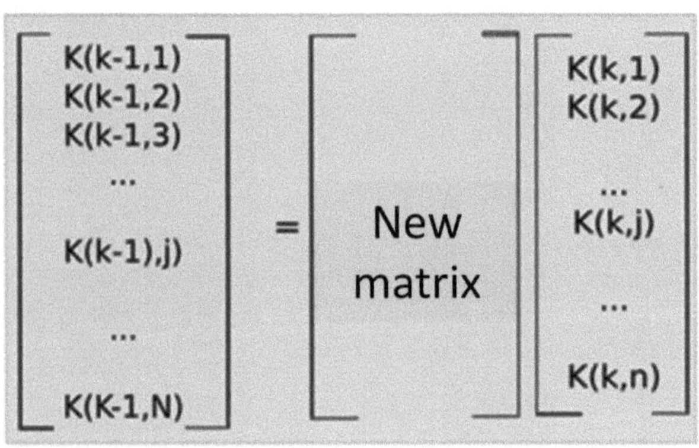

[2.130]

with:

K(k-1,1) = C(k-1,1)

K(k-1,2)= C(k-1,2) + [K(k-1,1)*A(k,2)] / U_1

K(k-1,j)=C(k-1,j) + [K(k-1,j-1)*A(k,j)] / Uj-1

...

K(k-1,n)=C(k-1,n) + [K(k-1,n-1)*A(k,n)] / Un-1 [2.131]

The general equation that translates the second matrix expression given may be written in the following form:

K(k-1,j)=Uj*C(k,j) – A(k,j+1)*C(k,j+1) [2.132]

with:

K(k-1,n)/Un

Thus, the matrix is in fact known at the instant k-1 and not at the instant k. In practice, we may "loop" by remaining at a fixed instant. Namely, the matrix is directly related to the diffusion coefficients which we know only at the instant n-1. After the first calculation of C(k,j), we recalculate the diffusion coefficients, and hence the matrix, while remaining at the initial point k-1 and thus until we satisfy a convergence criterion of the type:

$$\sum_{j=\&}^{n} \left| \frac{C(k,j)^p}{C(k,j)^{p-1}} \right| < \varepsilon \qquad [2.133]$$

where ε is generally equal for each (spatial) point j of the calculation.

NOTE.– Knowing that we often represent the logarithm for concentration, it would be judicious to carry out the variable change for: C= e^X, thereby resolving the diffusion equation at x.

2.11.1. *Matlab global code*

```
filename = 'fick.gif';
forkk=1:1:300
DopingBulk = 10^15;
    MaxDoping = 10^20;
Rp = 0;
DeltaRp = 5.0*10^(-9);
XSi = 5.0*10^(-8);
NT =201;
T         = kk ;
   DT = T/(NT-1);
C1 = zeros(1,NN);
NN = 101 ;
X         = zeros(1,NN);
X(1) = 0 ;
DX = (XSi)/(NN 1);
for j=2:NN
X(j)= X(j-1)+DX;
```

```
end
   C0=zeros(1,NN);
for j=1:NN
C0(j) = DopingBulk + MaxDoping / exp((X(j)-Rp)^2/DeltaRp^2);
end
   A0 = DT/(DX*DX);
   D0 = 4*10^(-19);
   DD = zeros(1,NN);
for j=1:NN
DD(j)= A0*D0;
end
for j=1:NN
C1(j)= C0(j);
end
   C2 = zeros(1,NN);
   U = zeros(1,NN);
   K = zeros(1,NN);
CONCENTRATION = zeros(NN,NT);
ALGORITHM from  FICK' LAWS
for n=1:NT
for j =1:NN
CONCENTRATION(j,n)= C1(j);
end
U(1)= 1.+DD(1);
for j=2:NN-1
U(j) = 1+(DD(j)+DD(j+1))-((DD(NN))^(2))/U(j-1);
end
      U(NN)=1+DD(NN)-((DD(NN))^2)/U(NN-1);
      K(1)=C1(1) ;
for j=2:NN
         K(j)=C1(j)+(K(j-1)*DD(j)/U(j-1));
end
C1(NN) = K(NN)/U(NN);
for j = NN:-1:2
         C1(j-1)= (K(j-1)+DD(j)*C1(j))/U(j-1);
end
for j=1:NN
         C2(j)= C1(j);
end
end
axex=zeros(1,NN);
for j =1:1:NN
axex(j)=j;
end
   plot(axex,C0,axex,C2);
```

```
ylim([0 10^20])
   M(kk)=getframe;
im = frame2im(M(kk));
   [imind,cm] = rgb2ind(im,256);
ifkk == 1;
imwrite(imind,cm,filename,'gif', 'Loopcount',inf,'DelayTime',0);
else
       imwrite(imind,cm,filename,'gif','WriteMode','append','DelayTime',0);
end
end
```

2.11.2. *Conclusion to Appendix 1*

In the field of microelectronics, the modeling of diffusion is essentially governed by the diffusion equation. However, what happens to this paradigm when we enter the field of nanoelectronics, for diffusion of dopants as well as for temperature, which is also governed by the diffusion equation or heat equation; what happens to the definition of temperature? We must certainly pay more attention to these phenomenological processes (see the Onsager theory) and stochastic processes. This would mean better introducing different microscopic processes for the diffusion of dopants, such as couplings with defects, which may be redhibitory, taking them into account in the equations.

2.12. Appendix 2

Ideal gas law: PV = nRT

2.12.1. *Introduction*

This small simulator (initiated in the AMERINSA foundation course at INSA and taken further by some departments) may make it possible to take into account a gas of particles in an enclosure, starting from the concepts of pressure and temperature.

This program might serve as a pre-processor to a Monte Carlo-type simulator, resolving the Boltzmann equation and bringing in, for example, the probabilities of electron–phonon interactions in a semiconductor (Si, GaAs) – refer to Chapter 3. Color versions of the figures in the following sections are available at www.iste.co.uk/gontrand/nanoelectronics.zip.

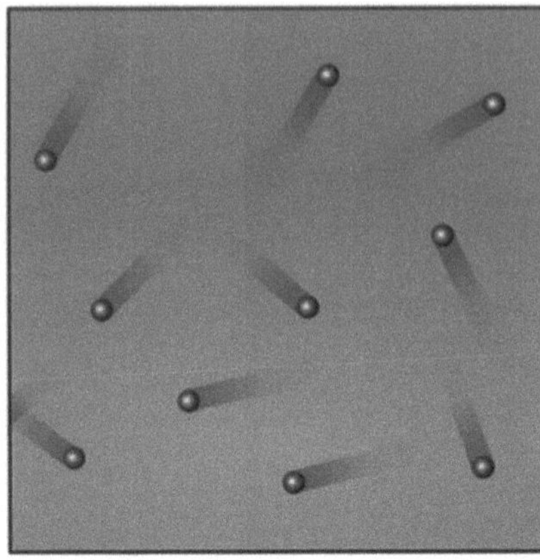

$$\vec{x}(t+\Delta t)+\vec{x}(t-\Delta t)=2\vec{x}(t)+\vec{a}(t)\Delta t^2+O(\Delta t^4)$$

The error committed with the Verlet method is of the order of Δt^4, *without any additional numerical effort!*

2.12.2. *Construction of a dynamic simulator*

We will now apply the Euler and Verlet resolution schema in order to construct a dynamic simulator. We will first apply this to a simple system (mass-spring) and then progressively complexify this in the following sections in order to arrive at the modeling for an ideal gas.

The following is the system to model:

A point mass m is situated at the end of a spring of stiffness k, which applies a force to the mass that is given by:

$$F_x = -k \cdot x$$

The variable x represents the position of the mass on the horizontal axis and we see that the position $x = 0$ corresponds to the equilibrium position for the spring (i.e. that for which it develops no stress). The mass is only free to move along the horizontal axis and its instantaneous speed is denoted by v_x. Starting from any initial conditions sx_0 and v_{x0}, we wish to put in place a simulation that allows us to foresee the trajectory of the mass over time:

– Here is a Matlab script titled "Masse_Ressort.m" (Mass-Spring): t_0 and t_f (initial and final instants of the simulation), Δt (time step at resolution), x_0 and v_{x0} (initial conditions), m and k.

```
%%%%%%%%%%%%%%%%%%%%%%%%%%%%%%%%%%%%%%%%%%%%%%%%%%%%%%%%%%%%%
%%%             MINI-PROJET DYNAMIQUE DE PARTICULES       %%%
%%%                  SYSTEME MASSE-RESSORT                %%%
%%%%%%%%%%%%%%%%%%%%%%%%%%%%%%%%%%%%%%%%%%%%%%%%%%%%%%%%%%%%%
clear all

% Données utilisateur :
T0=0;                       % Instant initial
Tf=20;                      % Instant final
dT=0.05;                    % Pas de temps
X0=1;                       % Position initiale
Vx0=0;                      % Vitesse initiale
M=1;                        % Masse
K=1;                        % Raideur
Integrator='Euler';         % Schéma : 'Verlet' ou 'Euler'

```

– Calculate the number of total steps in the simulation and create empty vectors (of the right size) to gather, over the evolution of the simulation, the variables t, x and v_x. Fill in t_0, x_0 and v_{x0} in the initial positions of these vectors.

Corrected Matlab code

```
% Nombre de pas :
Npas=(Tf-T0)/dT;

% Initialisation des instants :
T=[T0:dT:Tf]';

% Initialisation des positions :
X=zeros(Npas+1,1);
X(1)=X0;

% Initialisation des vitesses :
Vx=zeros(Npas+1,1);
Vx(1)=Vx0;

```

– Open the main loop. For each time step, calculate the value F_x. Then, apply Euler's schema to calculate the new values of x and v_x. Next, close the loop.

MATLAB code

```
% Boucle temporelle
for pas=1:Npas

    % Calcul de la force de rappel
    Fx=-K*X(pas);

    % Schéma
    if strcmp(Integrator,'Verlet')
        Vx(pas+1)=Vx(pas)+dT*Fx/M;
        X(pas+1)=X(pas)+dT*Vx(pas+1);
    elseif strcmp(Integrator,'Euler')
        X(pas+1)=X(pas)+dT*Vx(pas);
        Vx(pas+1)=Vx(pas)+dT*Fx/M;
    end
end

```

– Following the loop, based on the vectors that give the temporal evolution of x and v_x, calculate the temporal evolution of kinetic energy $m \cdot v_x^2/2$ and the potential energy $k \cdot x^2/2$, and then the total energy of the system.

– Trace the variables x and v_x as a function of time.

– Trace the kinetic, potential and total energies over time.

```

% Calcul des énergies cinétique, potentielle et totale
Ecx=0.5*M*Vx.^2;
Epx=0.5*K*X.^2;
Et=Ecx+Epx;

% Tracés
figure;plot(T,X,'.-b');title('Position')

figure;plot(T,Vx,'.-b');title('Vitesse')

figure
plot(T,Ecx,'.-b')
hold on
plot(T,Epx,'.-r')
plot(T,Et,'.-g')
title('Energies')
```

Choose a data set (e.g. $t_0 = 0$, $t_f = 20$, $\Delta t = 0.05$, $x_0 = 1$, $v_{x0} = 0$, $m = 1$ and $k = 1$). Launch the simulation. Is the energy report satisfactory? Conclusion?

Explicit non-conservative Euler scheme

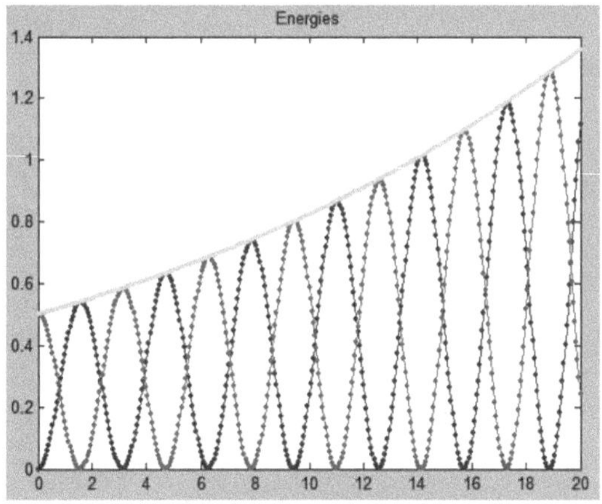

Contrary to an explicit Euler schema, via Verlet schema, the solutions are much better (numerical schema are a part and parcel of "numerical analysis" and thus mathematics, thereby falling outside our area of discussion here. It must be recognized that this domain is complex and open – see stability, convergence criteria).

This time the schema is conservative. The small oscillations observed in energy may easily be corrected using the "leap-frog" schema (see Chapter 2). Also refer to the FDTD (Finite Difference in Time Domain) method. The oscillations do not affect the stability of the solution.

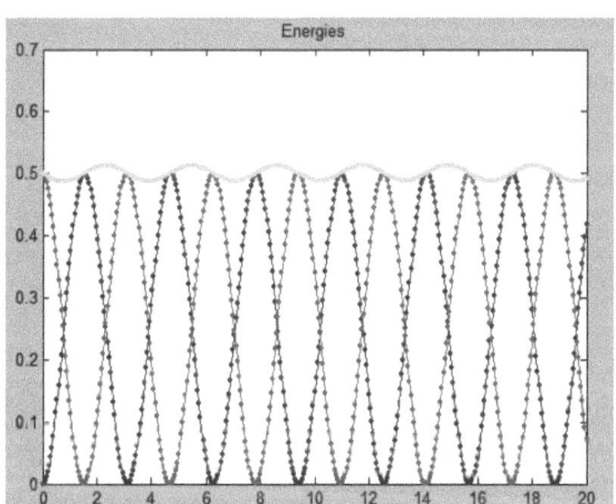

Launch several simulations with different time steps (going from 0.001 to 1). How does the solution evolve if we use the Euler scheme? And in the case of a Verlet scheme?

2.12.3. *Modeling the rebound of a particle on a wall*

In this section, we will make use of the earlier script to create a rebound simulator. The following is the system to model:

The mass is always considered to be a point mass (i.e. null dimensions); it is moved by an initial velocity v_{x0} (negative, if the rebound will not occur) and its position is situated with respect to that of the wall where the rebound will occur. This system is identical to the earlier system, except for the fact that the spring does not act unless the point mass penetrates into the wall (i.e. for $x < 0$). It can only push the mass from the wall, but cannot attract it in any case.

– Modify the initial conditions so as to make the rebound possible (i.e. impose $x_0 > 0$ and $v_{x0} < 0$).

Example of IC variables:

```
X0=5;                        % Position initiale
Vx0=-1;                      % Vitesse initiale
```

– Within the temporal loop, put in place a test at the time of the calculation of the retraction force: this force only acts when $x < 0$; otherwise, it is null.

Modified test loop with interpretation test

```
% Boucle temporelle
for pas=1:Npas

    % Calcul de la force de rappel
    if X(pas)<0
        Fx=-K*X(pas);
    else
        Fx=0;
    end

    % Schéma
    if strcmp(Integrator,'Verlet')
        Vx(pas+1)=Vx(pas)+dT*Fx/M;
        X(pas+1)=X(pas)+dT*Vx(pas+1);
    elseif strcmp(Integrator,'Euler')
        X(pas+1)=X(pas)+dT*Vx(pas);
        Vx(pas+1)=Vx(pas)+dT*Fx/M;
    end
end
```

Carry out these tests with finer or larger time steps.

Example of conservation of energy during the rebound when the Verlet schema is used:

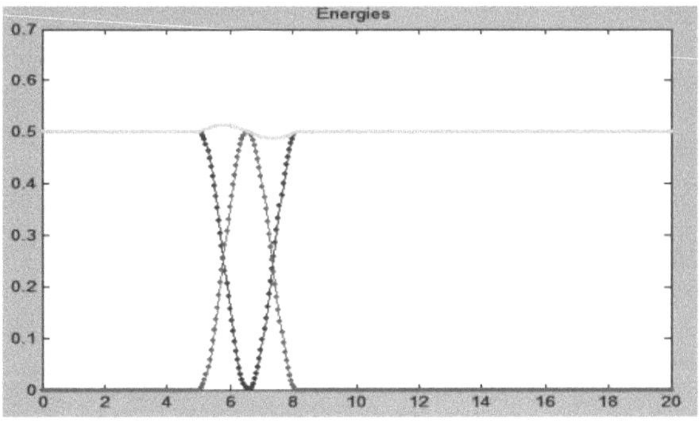

2.12.4. *Ideal gas in a box*

Here, we wish to apply the rebound model defined in the earlier section so as to experimentally find the ideal gas law using a simple numerical simulation. We will thus consider a point mass m in a rectangular box defined by the parameters x_{min}, x_{max}, y_{min} and y_{max}. The mass has the initial conditions x_0, y_0, v_{x0} and v_{y0}. The system has the following appearance:

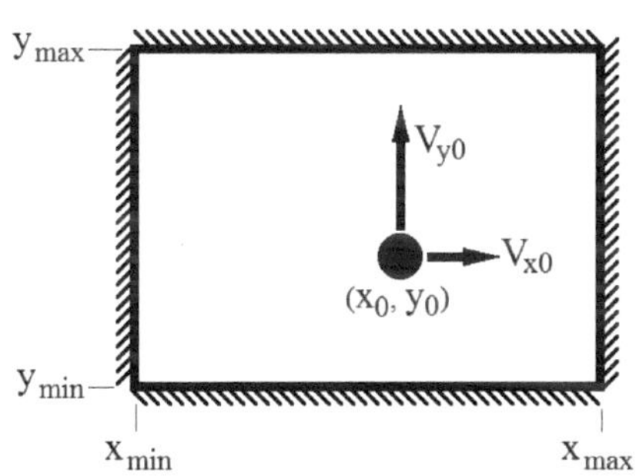

The difficulty here is to take into account the two degrees of freedom of the system.

– Make a copy of the "Rebound.m" script and rename it "Box.m".

– Complete the declaration of the variables using the new data.

```
%%%%%%%%%%%%%%%%%%%%%%%%%%%%%%%%%%%%%%%%%%%%%%%%%%%%%%%%%%%%%%%
%%%              MINI-PROJET DYNAMIQUE DE PARTICULES         %%%
%%%                            BOITE                          %%%
%%%%%%%%%%%%%%%%%%%%%%%%%%%%%%%%%%%%%%%%%%%%%%%%%%%%%%%%%%%%%%%
clear all

% Données utilisateur :
T0=0;                      % Instant initial
Tf=1000;                   % Instant final
dT=0.001;                  % Pas de temps
X0=10*rand;                % Position initiale en X
Y0=10*rand;                % Position initiale en Y
Vx0=rand;                  % Vitesse initiale en X
Vy0=rand;                  % Vitesse initiale en Y
Xmin=0;                    % Mur gauche
Xmax=10;                   % Mur droit
Ymin=0;                    % Mur inférieur
Ymax=10;                   % Mur supérieur
M=1;                       % Masse
K=1000;                    % Raideur
Integrator='Verlet';       % Schéma : 'Verlet' ou 'Euler'
```

– Before opening the new loop, create empty vectors of the right size to gather the temporal data on the vertical degrees of freedom (y and v_y). Do the same to store forces produced by the vertical (force F_x) and horizontal (force F_y) walls.

```
% Nombre de pas :
Npas=(Tf-T0)/dT;

% Initialisation des instants :
T=[T0:dT:Tf]';

% Initialisation des positions :
X=zeros(Npas+1,1);
X(1)=X0;
Y=zeros(Npas+1,1);
Y(1)=Y0;

% Initialisation des vitesses :
Vx=zeros(Npas+1,1);
Vx(1)=Vx0;
Vy=zeros(Npas+1,1);
Vy(1)=Vy0;

% Initialisation des forces :
Fx=zeros(Npas+1,1);
Fy=zeros(Npas+1,1);
```

– At the beginning of the loop, calculate the horizontal and vertical forces acting on the point mass at a given instant. This is non-null only if the mass goes out of the limits of the box.

Beginning of loop: tests for the calculation on the efforts on X and Y

```
% Boucle temporelle
for pas=1:Npas

    % Calcul de la force de rappel
    if X(pas)<Xmin
        Fx(pas)=-K*(X(pas)-Xmin);
    elseif X(pas)>Xmax
        Fx(pas)=-K*(X(pas)-Xmax);
    else
        Fx(pas)=0;
    end
    if Y(pas)<Ymin
        Fy(pas)=-K*(Y(pas)-Ymin);
    elseif Y(pas)>Ymax
        Fy(pas)=-K*(Y(pas)-Ymax);
    else
        Fy(pas)=0;
    end

```

– Implementation of the Verlet scheme for the vertical degree of freedom, in addition to the horizontal degree which must already be present.

End of loop: Verlet scheme for two limited developments

```
    % Schéma
    if strcmp(Integrator,'Verlet')
        Vx(pas+1)=Vx(pas)+dT*Fx(pas)/M;
        X(pas+1)=X(pas)+dT*Vx(pas+1);
        Vy(pas+1)=Vy(pas)+dT*Fy(pas)/M;
        Y(pas+1)=Y(pas)+dT*Vy(pas+1);
    elseif strcmp(Integrator,'Euler')
        X(pas+1)=X(pas)+dT*Vx(pas);
        Vx(pas+1)=Vx(pas)+dT*Fx(pas)/M;
        Y(pas+1)=Y(pas)+dT*Vy(pas);
        Vy(pas+1)=Vy(pas)+dT*Fy(pas)/M;
    end
end

```

– After exiting the loop, based on the temporal evolution of v_x, v_y, F_x and F_y, calculate the kinetic energy (along the two axes), potential energy of contact (along the two axes) and the total energy.

Calculation of kinetic and potential energies (a test is necessary for the potential energy at contacts)

```
% Calcul des énergies cinétique, potentielle et totale
Ecx=0.5*M*Vx.^2;
Ecy=0.5*M*Vy.^2;
Epx=zeros(Npas+1,1);
Epy=zeros(Npas+1,1);
for pas=1:Npas
    if X(pas)<Xmin
        Epx(pas)=0.5*K*(X(pas)-Xmin)^2;
    elseif X(pas)>Xmax
        Epx(pas)=0.5*K*(Xmax-X(pas))^2;
    end
    if Y(pas)<Ymin
        Epy(pas)=0.5*K*(Y(pas)-Ymin)^2;
    elseif Y(pas)>Ymax
        Epy(pas)=0.5*K*(Ymax-Y(pas))^2;
    end
end
Et=Ecx+Ecy+Epx+Epy;

```

– From this data, calculate the scalars P (pressure) and T (temperature), averaged over the duration of the simulation. Also calculate the volume, V, of the box and the scalar PV/T. Display the results.

Calculation of thermodynamic functions from energies. We find PV/T very closed from 1 if the simulation process is long enough.

```
% Calcul de la pression et de la température
disp(' ')
Pression=(mean(abs(Fx))+mean(abs(Fy)))/((Xmax-Xmin)*2+(Ymax-Ymin)*2);
disp(['Pression=',num2str(Pression)])
Temperature=mean(Ecx+Ecy);
disp(['Temperature=',num2str(Temperature)])
PVsurNT=Pression*(Xmax-Xmin)*(Ymax-Ymin)/Temperature;
disp(['PV/NT=',num2str(PVsurNT)])

```

The simulator must be turned on over a duration that is long enough for the time averages to have a direction (see ergodicity: the mean of the sample is equal to the time average) with different initial conditions.

An example of trajectory result.

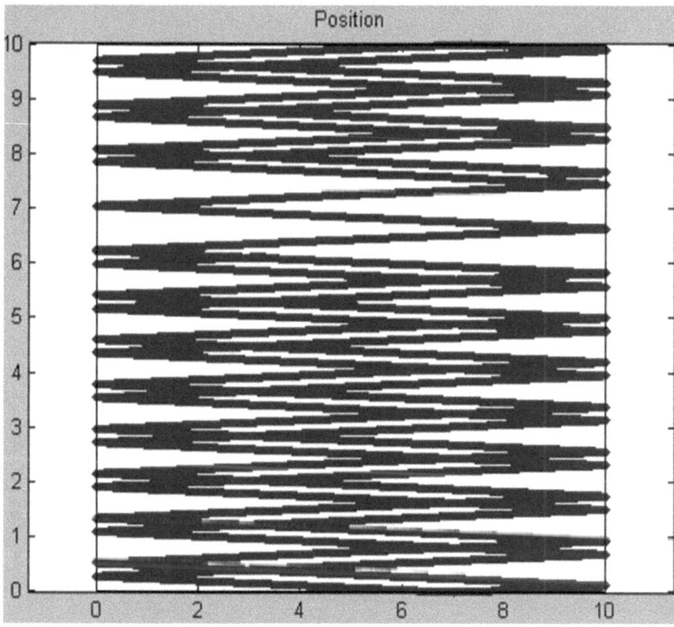

3

Electrical Functioning of Devices

SUMMARY.– For the past 40 years, the drift diffusion model has been at the heart of the electrical modeling of devices in terms of microelectronics; divergence of current density and "Poisson" are coupled equations. However, for 30 years now, devices have been getting smaller and smaller, and expert physicists and electronics engineers have been interested in backward master equations that are more fundamental like the Boltzmann equation. For 20 years now, eminent mathematicians have been interested in these integro-differential equations, looking for a possible bridge between the two. Furthermore, quantum mechanics has become increasingly pertinent. In fact, we are entering nanoelectronics, which should prompt researchers to come up with various paradigms in practice and partial or whole new sets of equations.

3.1. Introduction

At present, the underlying models of electrical simulations (device) are mainly two-dimensional and concern the DDM (drift diffusion model), relevant currents being induced by carrier drift (mainly under an electric field), electrons or holes, with concurrent diffusion of carrier packets.

3.1.1. *Near-future directions*

In this case, the advent of submicronic devices has rendered one-dimensional numerical models obsolete, as physical and geometric effects have to be taken into account together.

Numerical simulations and experiments need to match compact electrical models (which manufacturers are fond of) based on Kirchhoff equations for the quick resolution of large equation systems, while ensuring accuracy.

One scientific motivation is a relatively recent development in microelectronics. In this case, we are shifting from the latter to nanoelectronics, at least in some domains.

Moreover, why not use IV–IV devices: this would mean focusing on very small dimensions (<25 nm), as well as low-dimensionality systems, which explains new transport phenomena having first-order repercussions on transport, for instance the ultimate CMOS, and a desire to capitalize on them for new generations of nanoelectronics (Coulomb blockade, wave propagation effects):

– on ultra-thin, possibly strained, films (quantum confinement);

– with ultra-short gate and hence 2D effects and non-equilibrium transport (ballistic limit);

– with a possible coupling effect or resonant tunneling;

– with the Coulomb blockade effect on quantum islands.

A new physical phenomenon should be taken into account: unsteady quantum. However, what is lacking are robust simulation tools that take into account all of these effects. This is a real barrier to optimization and, even worse, to the evaluation of the potential of unconventional architecture or devices.

The complementarities of software-based approaches in technological, mathematical or other associated skills must also be highlighted.

The types of devices that we need to simulate are:

– In 1 or 2D: MOS capacitor (with quantum confinement and tunneling leakage via gate dielectrics), resonant-tunneling diodes and heterojunctions.

– In 2D: MOS with FD-SOI variant (Fully Depleted Silicon On Insulator), dual gate, plus material engineering in two directions (channels along the heterojunctions and also S/D (source/drain) using an SC (semiconductor) different from that used in the channel). Can we consider a tunneling effect through the gate and in the channel direction?

– In 3D: nodules (or nanocrystals, etc.).

– What is a possible and desirable development, and what is realistic, interesting and unnatural for simulation of unsteady quantum transport in some geometric structures?

– Monte Carlo modeling for simulation of semiconductor devices in one, two or three dimensions – an approach to transport that is conventional (see Boltzmann equation [CER 88]), but allows for taking into account carriers/phonons scatterings and describing unsteady phenomena. No ripple effect in the transport direction, but tunneling leakage through the gate oxide. Introduction of the vertical quantum confinement effect on transport. Evident complementarity with ballistic quantum approaches that, by definition, neglect interactions but describe all ripple effects (*a priori*).

– Poisson/Schrödinger modeling, 1D or 2D, MOS structures (well advanced), possibly considering heterojunctions for calculating charge control and tunneling current. The approach developed for calculating tunneling current (determining life duration of quasi-linked states in the channel) could be used to study resonant-tunneling diodes.

– Charge effects on semiconductor quantum dots and physical simulation of transport on structures containing such dots where a Coulomb blockade is possible. For now, (1) we know how to calculate energy levels in a 3D dot ("Poisson/Schrödinger" – auto-coherent) based on the size of the island, the number of electrons it has and the polarization, and (2) we use different approaches, in particular "Hartree–Fock" and "density functional theory" with approximation for the effective mass, as well as strong bonds to compare with the effective mass. Modeling current with a dot-having structure, the final objective being a multi-dot structure (with a more physical description than common phenomenological approaches with capacitors and resistors).

Observation: even here, as with diffusion and Brownian motion, random microscopic phenomena can induce deterministic macroscopic behavior: for instance, the saturation velocity of silicon is typically (low doping) 10^5 at ambient temperature, while it is inferred to as a statistical average (via a Monte Carlo algorithm simulating random carrier scattering on the network).

3.1.2. *Scientific motivation*

Reducing sizes will inevitably call into question quasi-conventional modeling. If quantum aspects seem to be inevitable here, they are often deliberately sought ("quantum devices").

Examples of devices that can be modeled in only one dimension: resonant-tunneling diodes, photo-devices, quantum cascade laser and superlattice.

In general, the devices used are quantum wires or dots. Currently, there are no simulation tools that are rigorous and general enough, which could be counterparts of the Monte Carlo methods or "dynamic" for conventional devices. Nevertheless, there are models for "ballistic" devices. When wave phenomena and dissipation by interactions are both far from negligible, there is no method that is really satisfying.

Physical simulation of quantum devices takes into account both wave effects and the interactions with the environment.

For the suggested contribution, the Monte Carlo method must be used. Moreover, with the latter, it is relatively easy to change from 2D to 3D. In fact, projecting the wave vector or the velocity – which allows for more research on electromagnetic phenomena – via Lorentz force ($q\vec{v}\wedge\vec{B}$), for instance on three axes, is not as complex as on two axes. There are more promising approaches, in particular Green's function formalisms, which are currently in vogue. It would be interesting to evaluate and compare such methods.

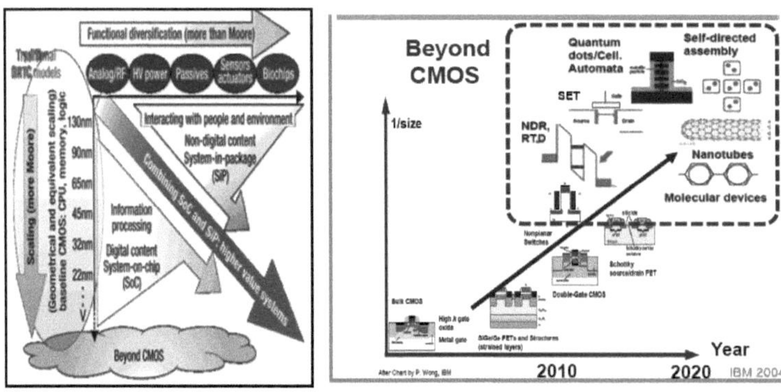

Figure 3.1. *Evolution of device technology at nanoscale (by P. Wong, IBM). For a color version of this figure, see www.iste.co.uk/gontrand/nanoelectronics.zip*

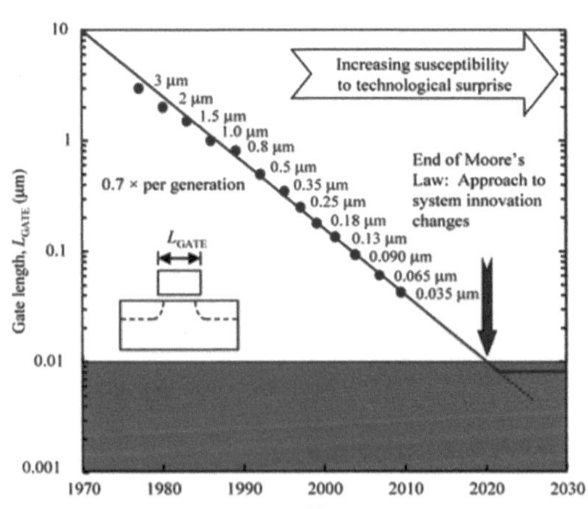

Figure 3.2. *Evolution of the length of a critical element of MOS: gate. For a color version of this figure, see www.iste.co.uk/gontrand/nanoelectronics.zip*

3.1.3. *Current*

3.1.3.1. *Monte Carlo electrical simulation [MOG 90]*

This simulation treats electron and hole movements in a semi-conventional manner, not taking into account quantization effects. Initially, it was envisioned for HEMT (High Electron Mobility Transistor; formerly TEGFET: Two-dimensional Electron Gas Field-Effect Transistor) devices, but now it can be applied to a wide range of devices for practically all possible III-V material systems. It describes a realistic electrode geometry, taking into account the presence of dielectrics, surface potential, limitation of pseudomorphic material and polarizing charge in nitrides of hexagonal structure, impact ionization and so on. Special efforts are put into modeling ohmic contacts.

Potential future development includes adaptation to conventional devices; these simulators cannot directly deal with the most fundamental quantum transport aspects. However, some phenomena can be introduced approximately: for example, tunneling currents through gates or heterojunctions, or effects linked with discrete doping distribution. This will make it possible to extend the validity of modeling for low-dimensionality devices, with no adverse effects, and provide a base for comparison for more rigorous studies.

3.1.3.2. *Transport simulation in heterostructures*

This simulation takes into account the quantization of electronic states (see Schrödinger/Poisson). It is being developed for III-V material, but is generalized. It takes into account the quantization of all valleys; their parabolicity (see wave vector/energy relation) has to be described rigorously. There has been significant work on applying this to the case of high electron density (highly degenerate gas), and it must be extended to the study of background noise phenomena.

There are software that solve 1D Schrödinger's equation. They were developed for "parallel" transport, that is, in the direction of growth; however, generalization in the case of "vertical" transport, perpendicular to the interfaces, is on the right track.

3.1.4. *Possible approach*

We take as a starting point the "Pauli master equation" method (see also the "Fokker–Planck" equation). We can consider it as a generalization of the Boltzmann equation and use a Monte Carlo method. Let us note that the "Monte Carlo" method gives an exact solution of this latter equation, although it is based on statistical methods.

Electronic dynamics is described in terms of jumps between quantum states, including "transient" states that transport current. The "accelerator" field acts only through the Hamiltonian potential that helps determine electronic states (value = eigenvalue).

Ab initio, this method only takes into account the diagonal elements of the density matrix, ignoring coherent phenomena. However, these can be taken into account via a "generalized Monte Carlo" method. Generally, this approach runs the risk of leading to very heavy calculations. We must generalize, moreover, to various geometrical forms, 2D or 3D.

It must also be highlighted that most studies appearing in the literature use "strong approximation" for mechanisms of interaction and band structure of material. These approximations will undoubtedly be unacceptable for even remotely realistic simulations. In some cases, simple approximations are possible and in others, a fully sophisticated numerical approach can be indispensable (see k.p. theory); undoubtedly, this will be the case for valence bands. *A priori*, a Hamiltonian with a sufficient number of bands will be enough, but there are other formulations, such as Green's kernel approaches.

3.2. Current state diagram

3.2.1. *Drift diffusion model*

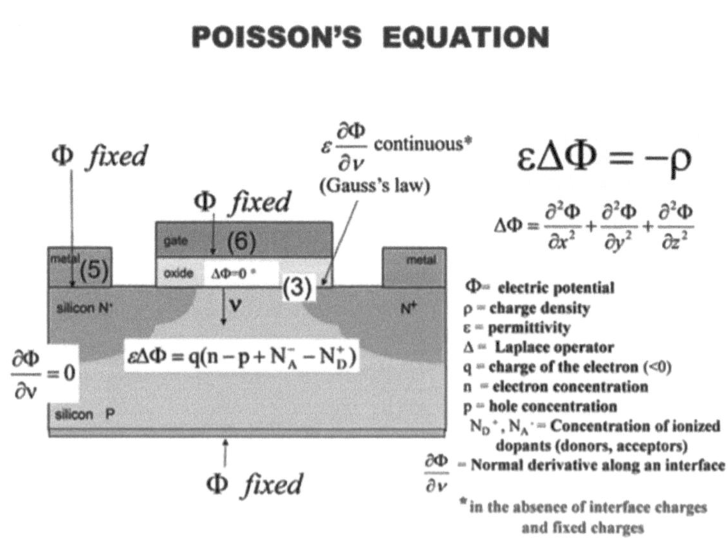

Figure 3.3. *Poisson's equation applied to the MOSFET. For a color version of this figure, see www.iste.co.uk/gontrand/nanoelectronics.zip*

DRIFT DIFFUSION MODEL

In each semiconductor

- Poisson equation for electrostatic potential

$$\varepsilon\Delta\Phi = q(n - p + N_A^- - N_D^+)$$

- current = diffusion current + drift current
  - action of electric field $\vec{E} = -\vec{\nabla}\Phi$
  - D_n, D_p = diffusivity, μ_n, μ_p = mobility

$$\overrightarrow{J_n} = -q\left(n\mu_n\vec{\nabla}\Phi - D_n\vec{\nabla}n\right)$$

$$\overrightarrow{J_p} = -q\left(p\mu_p\vec{\nabla}\Phi + D_p\vec{\nabla}p\right)$$

- charge balance
  - $\nabla\left(\overrightarrow{J_n} + \overrightarrow{J_p}\right) + q\frac{\partial}{\partial t}(p - n)) = 0$

$$\frac{\partial n}{\partial t} - \nabla\left(\frac{\overrightarrow{J_n}}{q}\right) = G - R$$

$$\frac{\partial p}{\partial t} + \nabla\left(\frac{\overrightarrow{J_p}}{q}\right) = G - R$$

- steady state and without generation-recombination

$$\nabla\overrightarrow{J_n} = \nabla\overrightarrow{J_p} = 0$$

Figure 3.4. *The drift diffusion model*

... INTRODUCTION TO MORE PHYSICAL STUDIES...

Poisson and drift-diffusion equations are formulations that are independent of the nature of materials

... **where do the semiconductor properties of silicon play a role?**

Answer ... in boundary and interface conditions
Which constants?
What happens in a nonuniformely doped semiconductoor's voulume, in xin which all the contacts are grounded

The current is null, but the electric potential cannot be constant because an elctric fieds stays around the junctions
⇒2 elecrochimical potentials appear, called pseudo levels of Fermi (n and p), which will be constant when no current is present, and related later to the statistics of occupation of the energy levels of the charges carriers
⇒Value of polaryzarion applied extracted from solutions with constant Fermi levels.

Figure 3.5. *Local conditions*

BOLTZMANN / POISSON STATISTICS

$$n = n_i\, e^{\frac{\Phi - \phi_n}{U_T}} \qquad p = n_i\, e^{\frac{\phi_p - \Phi}{U_T}}$$

$$\varepsilon\Delta\Phi = q\left(n_i e^{\frac{\Phi - \phi_n}{U_T}} - n_i e^{\frac{\psi_p - \Phi}{U_T}} + N_A^- - N_D^+\right)$$

Φ= electric potential
ϕ_n, ϕ_n quasi Fermi levels

n_i = intrinsic Concentration
$U_T = k_B T/q$
k_B = Boltzmann constant

Figure 3.6. *Boltzmann's statistics for Poisson's equation*

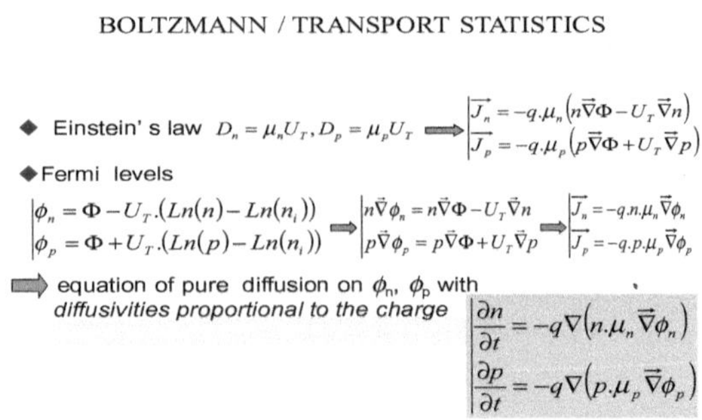

Figure 3.7. *Botzmann's statistics for transport equations*

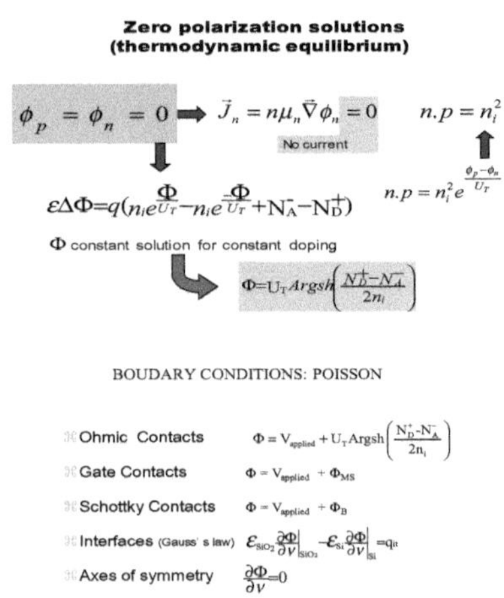

Figure 3.8. *Potential at thermodynamic equilibrium extracted from Poisson's equation. For a color version of this figure, see www.iste.co.uk/gontrand/nanoelectronics.zip*

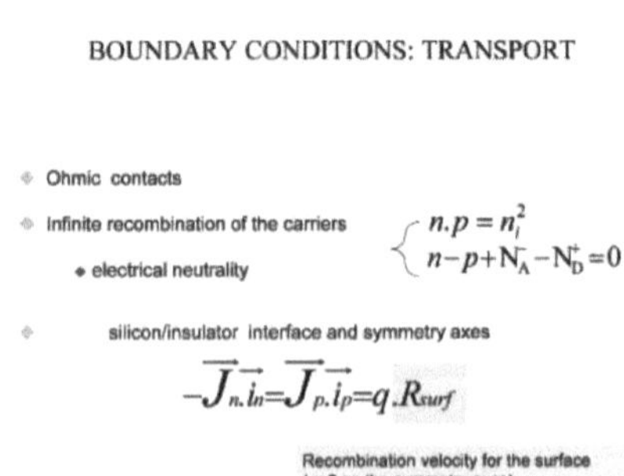

Figure 3.9. *Transport: boundary conditions*

DISCRETIZATION OF THE LAPLACE OPERATOR

For a second derivarive approach:

1st order schematic, centered constant mesh: $h=x_{i+1}-x_i$ for any i

$$f(x_{i+1}) = f(x_i) + h.f'(x_i) + \frac{1}{2!}h^2 f''(x_i) + \frac{1}{3!}h^3 f'''(\xi_+)$$

$$f(x_{i-1}) = f(x_i) - h.f'(x_i) + \frac{1}{2!}h^2 f''(x_i) - \frac{1}{3!}h^3 f'''(\xi_-)$$

$$f''(x_i) = \frac{f(x_{i+1}) + f(x_{i-1}) - 2f(x_i)}{h^2} - \frac{1}{6}h\left(f'''(\xi_+) - f'''(\xi_-)\right)$$

Same for the other space variable (2D, 3D)

Figure 3.10. *Laplacian discretization*

FINITE DIFFERENCES METHOD

Principles

- ◆ applying Taylor formula in each spatial dimension

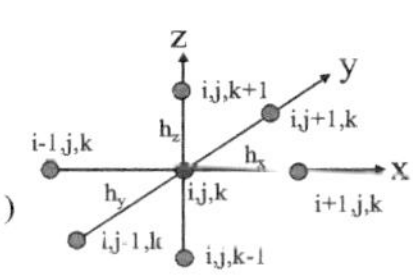

$$\Delta\Phi(x_i, y_j, z_k) - \Delta_{ijk}(\Phi) + h_x\varepsilon(h_x) + h_y\varepsilon(h_y) + h_z\varepsilon(h_z)$$

$\varepsilon(h_x)$, $\varepsilon(h_y)$ $\varepsilon(h_z)$ tends to 0 while h_x, h_y and h_z tend to 0 .

$$\Delta_{ijk}(\Phi) = \frac{\Phi_{i+1jk} + \Phi_{i-1jk} - 2\Phi_{ijk}}{h_x^2} + \frac{\Phi_{ij+1k} + \Phi_{ij-1k} - 2\Phi_{ijk}}{h_y^2} + \frac{\Phi_{ijk+1} + \Phi_{ijk-1} - 2\Phi_{ijk}}{h_z^2}$$

- ◆ examples boundary condition: Φ_{ijk} fixed ; $\frac{\Phi_{i+1jk} - \Phi_{ijk}}{h_x} = 0$
- ◆ solution of a non-linear system ➡ Method of' Newton-Raphson

➡ Linear system with sparse matrix

Figure 3.11. *Finite differences method. For a color version of this figure, see www.iste.co.uk/gontrand/nanoelectronics.zip*

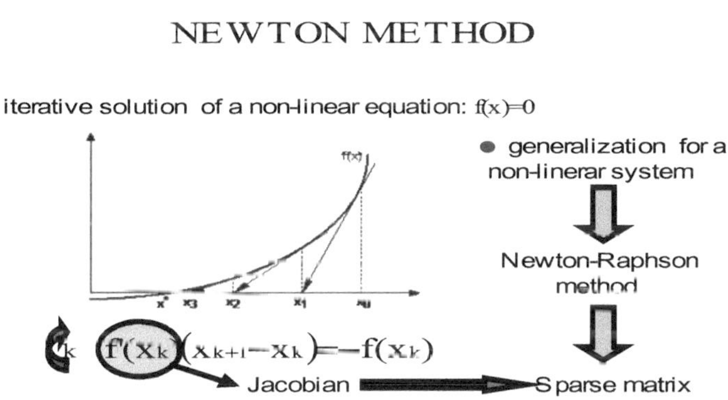

Figure 3.12. *Newton method. For a color version of this figure, see www.iste.co.uk/gontrand/nanoelectronics.zip*

NEWTON-RAPHSON METHOD
FOR POISSON'S EQUATION

$$\varepsilon.f''-a.e^{f}=0$$

With the finite difference in 1D:

$$f''(x_i)\approx\frac{f(x_{i+1})+f(x_{i-1})-2f(x_i)}{h^2}$$

$$\Rightarrow\ \varepsilon\left(f_{i+1}+f_{i-1}-2f_i\right)-a.h^2e^{f_i}=F(f_{i-1},f_i,f_{i+1})=0$$

$$\Rightarrow\ \frac{\partial F}{\partial f_{i-1}}\left(f_{i-1}^{k+1}-f_{i-1}^{k}\right)+\frac{\partial F}{\partial f_i}\left(f_i^{k+1}-f_i^{k}\right)+\frac{\partial F}{\partial f_{i+1}}\left(f_{i+1}^{k+1}-f_{i+1}^{k}\right)=-F\left(f_{i-1}^{k},f_i^{k},f_{i+1}^{k}\right)\ \circlearrowleft k$$

$$\left(-2\varepsilon-a.h^2e^{f_i}\right)\left(f_i^{k+1}-f_i^{k}\right)+\varepsilon\left(f_{i+1}^{k+1}-f_{i+1}^{k}\right)+\varepsilon\left(f_{i-1}^{k+1}-f_{i-1}^{k}\right)=a.h^2.e^{f_i}+\varepsilon\left(2f_i^{k}-f_{i+1}^{k}-f_{i-1}^{k}\right)$$

→linear equation in the $\left(f_i^{k+1}-f_i^{k}\right)$
→for all nodes i of the mesh

Figure 3.13. *Newton method for Poisson's equation*

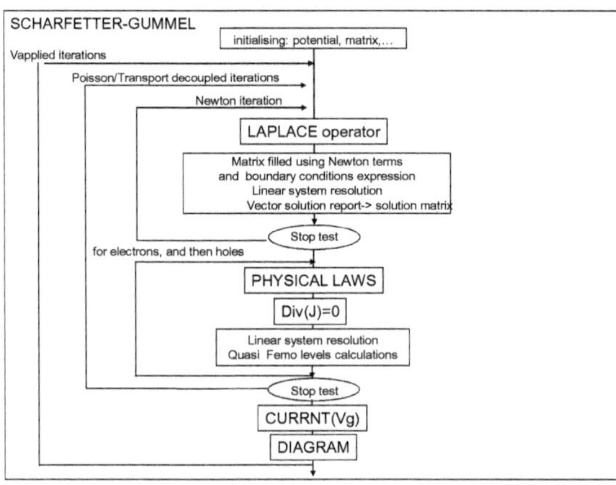

Figure 3.14. *Scharfetter–Gummel's flow chart*

DISCRETIZATION OF DRIFT DIFFUSION EQUATIONS

⌘ **NONOPERATIVE** centered schemas
(finite differences and finite elements
c.f. Poisson's equation or pure diffusion)

⇒ Risks of oscillations resulting from
- Very coarse meshes
- $\vec{E}$ too strong

⇒ upstream decentering (finite differences) or
local conservation of current (finite volumes)

Figure 3.15. *Discretization of drift diffusion equations. For a color version of this figure, see www.iste.co.uk/gontrand/nanoelectronics.zip*

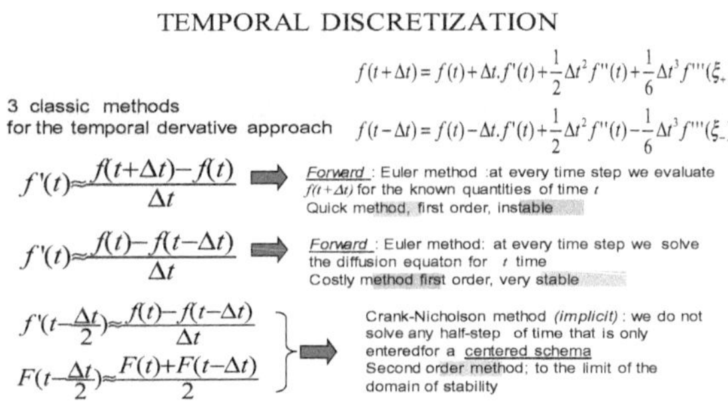

Figure 3.16. *Temporal discretization. For a color version of this figure, see www.iste.co.uk/gontrand/nanoelectronics.zip*

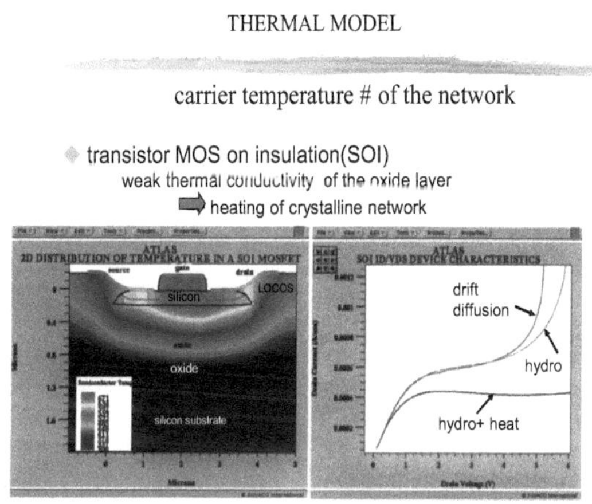

Figure 3.17. *MOS; 2D simulations: temperature map and I(V). SILVACO. For a color version of this figure, see www.iste.co.uk/gontrand/nanoelectronics.zip*

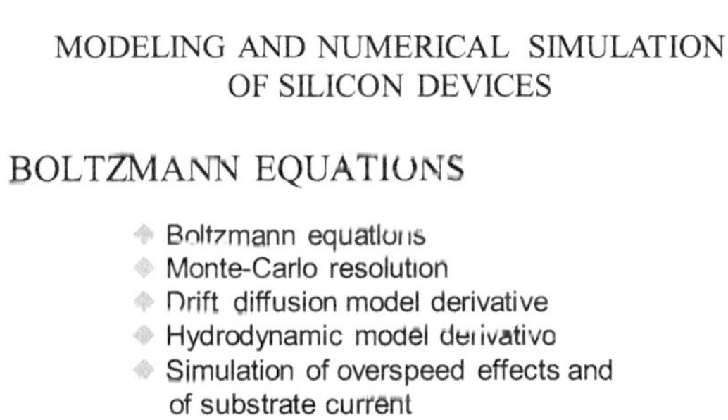

Figure 3.18. *From Boltzmann's equation*

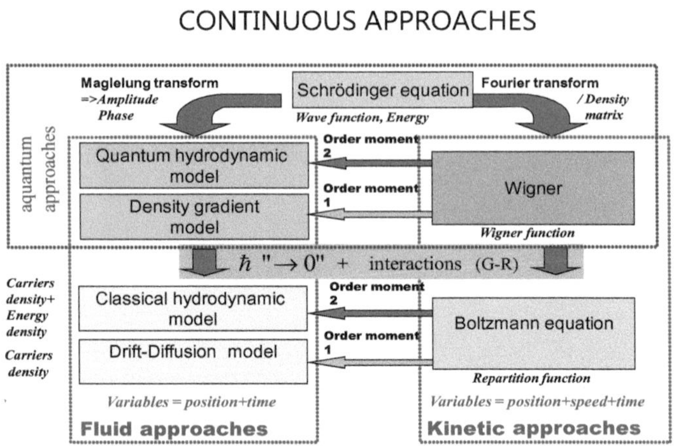

Figure 3.19. *Continuous equations for device modeling. For a color version of this figure, see www.iste.co.uk/gontrand/nanoelectronics.zip*

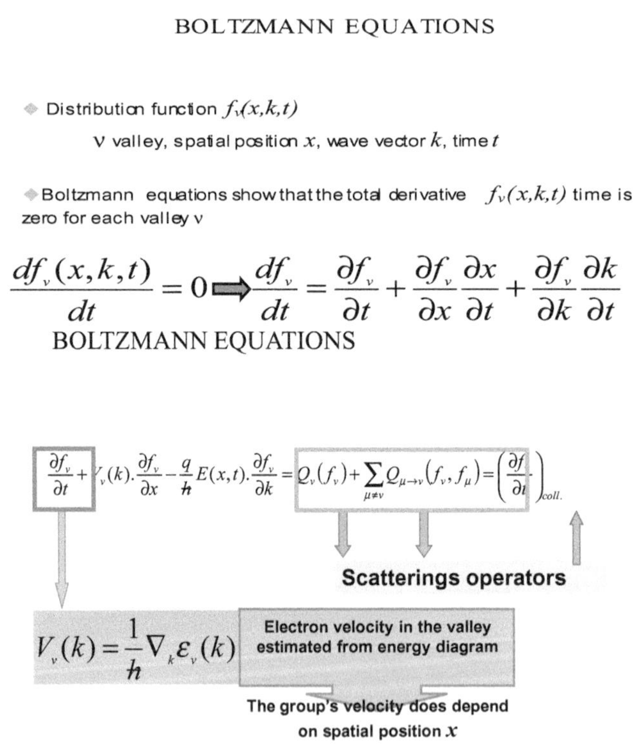

Figure 3.20. *Boltzmann equations. For a color version of this figure, see www.iste.co.uk/gontrand/nanoelectronics.zip*

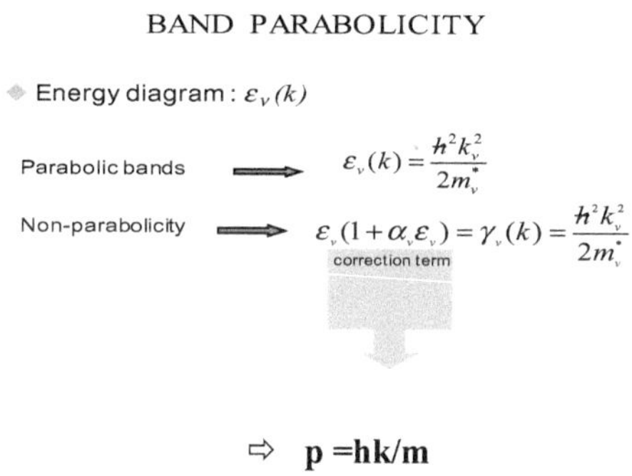

Figure 3.21. *Band parabolicity*

3.2.1.1. *On Boltzmann equation [4.5], see Figure 3.20*

When solving partial derivation equations, we start with complex models and simplify the equations in certain cases. In fact, it is unacceptable to apply laws of physics to all particles, such as those of mechanics, taking into account positions and velocities because there will be too much freedom and too many equations.

For example, to simplify, we take only velocities into account and arrive at the results using, for instance, Navier–Stokes equations, at least for some of these reductions. Mathematicians were able to arrive at the heat equation using the linear Boltzmann equation, which is itself a master equation for the kinetic theory of collisions.

The hypothesis, among others, is:

– particles are indiscernible;

– collision time is zero (hard spheres, elastic shocks).

Boltzmann equation

– one expression of the equation:

$$\frac{\partial f_\nu}{\partial t}+V_\nu(k).\frac{\partial f_\nu}{\partial x}-\frac{q}{\hbar}E(x,t).\frac{\partial f_\nu}{\partial k}=Q_\nu\left(f_\nu\right)+\sum_{\mu\neq\nu}Q_{\mu\to\nu}\left(f_\nu,f_\mu\right)=\left(\frac{\partial f_\nu}{\partial t}\right)_{coll.}$$

Let us consider a gas composed of n particles, where $10^{22} > n > 10^6$.

A particle can represent the statistical mean of a certain number of carriers, for example electrical ones. A microscopic study of Boltzmann equation, representing the probability density of equations, is carried out in three phases: three dimensions of positions x, y, z, and three velocities v_x, v_y, v_z or wave vectors (m^{-1}) k_x, ky, kz (in the case of a parabolic energy band – $\varepsilon = h^2k^2/2m$: $v = hk/m$). The whole set of particles is similar to a gas that is rarified "enough" to assume that at least three particles will not interact simultaneously.

The second term on the left, $v.\vec{\nabla}_X.f$ for transport, shows that the particles travel in a straight line.

Term Q: for collision (scattering) is input–output, on one state of the wave vector k (m^{-1}).

Rq: during numerical calculation, it might prove useful to keep the output Q in its integral form, and not calculate its integral, numerical errors on two inputs/outputs tending to offset each other.

Here, we will only mention that it is possible to easily solve Shur equations, simplification of the Boltzmann model, but it is necessary to know the steady state.

Hydrodynamic model

The fluid is assumed to be incompressible. In this case, the fluid density is constant, which means that there is zero divergence of the velocity field of the set of particles concerned, occupying the same volume before and after the shock; this is not truly compatible when coupled with the Poisson equation, which changes attraction or repulsion between carriers in a volume element.

In fact, a kinetic approach is more complex than a hydrodynamic one, as the work is carried out in the phase space; input for the collision integral is also time-consuming during calculation.

When switching from Boltzmann kinetics to hydrodynamic models (asymptotic?), thermodynamic equilibrium, even local, is assumed to exist. In practice, the collision operator does not cause variations to the Boltzmann distribution f; on average, input equals output, on state k, which implies a Maxwellian form of f, associated with maximum entropy (see entropic distance), locally (see Gibbs lemma). For small disturbances, at weak electric fields, for instance, solutions for f can still be considered, like the Maxwell distribution, but along the whole length of the field.

Moreover, can we deduce fluidic equations from Newtonian mechanics?

Hilbert proposed to go via the Boltzmann equation; this would restrict *de facto* possible solutions, in particular, because of the introduction of an ideal gas model.

The Boltzmann equation can also be considered to be an approximation of the compressible Euler equation for a small Knudsen number (linked to the ratio of the mean free path to the nominal size of the concerned system).

From the Boltzmann equation, we can also deduce some forms for Navier–Stokes equations (for some heat laws).

Ignorance of compressible equations

A compressible Navier–Stokes equation cannot be *a priori* deduced from the Boltzmann equation; however, it can be deduced from a compressible Euler equation, another master equation.

If fluctuations of f around equilibrium with respect to the Navier–Stokes equation seem to be very difficult to understand, then what about chaos or shockwave modeling?

Either way, much mathematical work of the highest quality is in full swing, in particular with respect to the asymptotic expansion of such equations.

Problems linked to "device modeling" are *a priori* less complex for solving parabolic diffusion equations than for solving the nonlinear elliptic Poisson equation or low levels of applied voltage, where we often use quasi-Fermi levels. The three equations, Poisson's, divergence of electron and hole current density, are linear elliptic.

However, if we define, for instance, Schottky contacts in exponential form – exp(qV/kT) – then related equations of current density divergence will become linear elliptic; underflow or overflows may seem to be linked to these exponentials.

However, in the case of a strong voltage or electric field, there will be a strong coupling between the three equations. Terms referring to the drift current due to the electric field will become preponderant, and quasi-Fermi level limits may lead to numerical problems. As for the displacement current term, which is more important when at high frequency, it is sometimes preferable to keep it, even at low frequency, for numerical stability reasons. Each equation is linear in the main variable.

If we do not introduce generation–recombination terms, equations related to electrons and holes lead to an asymmetric system, for which the solution is more complex than for symmetric systems.

Several models have been derived based on the Boltzmann equation. Statistical methods like Monte Carlo seem to be more suitable for simulation of submicronic devices; what's more, the smaller the devices, the lesser they will be time-consuming in terms of calculation, the limit being: one particle < --> one carrier.

When we get into the submicronic field, Monte Carlo methods seem more interesting, coupled with the Coulomb law, to take into account the interaction between carriers, thus dropping the Poisson equation. In a given grid, we can calculate distances between barycenters of meshes taken 2 by 2 and stock them at the beginning of the program; then, it suffices to upcount the particles that are entering and down count the particles that are exiting, to adjust to each (quite a few) fixed time step(s) and readjust the electric field in the concerned meshes.

The velocity autocorrelation function, easy to introduce in these statistical algorithms, must be the basic tool used in analyzing diffusion noise, while also taking into account the network phonon-carrier scattering time. Even here, the duration of shocks and phonon exchange for inelastic interactions must take into account the Heisenberg paradigm ($\Delta\varepsilon\Delta t > h/2/\pi$).

MEAN VALUES

Carrier densitiy electrons

$$n(x,t) = \sum_{v} n_v(x,t) = \sum_{v} \int_{R^3} \frac{f_v(x,k,t)Z_v}{4\pi^3} dk$$

Carrier mean velocity

$$\langle V \rangle(x,t) = \frac{1}{n(x,t)} \sum_{v} \int_{R^3} V_v(k) \frac{f_v(x,k,t)Z_v}{4\pi^3} dk$$

Carrier mean energy

$$\langle \varepsilon \rangle(x,t) = \frac{1}{n(x,t)} \sum_{v} \int_{R^3} \varepsilon_v(k) \frac{f_v(x,k,t)Z_v}{4\pi^3} dk$$

Figure 3.22. *Mean values of physical variables*

NUMERICAL SOLUTIONS OF BOLTZMANN EQUATION

Collision terms make Boltzmann's equation into integro-differential equations function of the wave vector k

It is hard to imagine numerical solutions of Boltzmann equations using conventional discretization methods (counter example: Taylor development to the fourth or fifth order).

→ Deterministic or stochastic particle methods

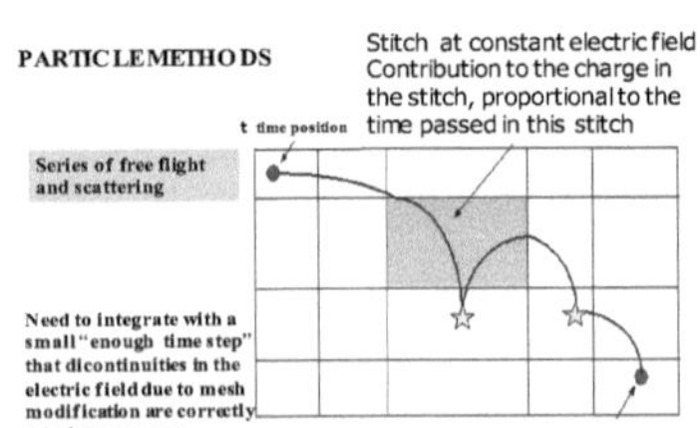

Figure 3.23. *Stochastic particle methods*

Succession of free flights and scatterings

MONTE CARLO METHOD

Determination of the free light time *tv*

- Progression in time
- Choice of an interaction
- Selected scattering effect
- Screening

Figure 3.24. *Monte Carlo method*

2/ Progress over time

- After free flight for time t_v, position x and wave vector k are:

$$x = x_0 + \frac{\hbar k_0 t}{m^*} + \frac{q}{m^*}\int_0^{t_v}\left(\int_0^t E(\tau)d\tau\right)dt$$

$$k = k_0 + \frac{q}{\hbar}\int_0^{t_v} E(t)dt$$

- Need to integrate with sufficiently small time steps so that discontinuities of the electric field E due to mesh changes are correctly taken into account. (and also, to have "enough" free flights as compared to interactions with the network (cf. Poisson process)

Figure 3.25. *Progress over time for the wave vector*

3/ Scattering choice

- We assume that interactions are independent between themselves in a time step.

- We draw lots a number r between zero and λ_{max}
- = Interaction K is defined as :

$$\sum_{i=1}^{k-1}\lambda_i(\varepsilon) < r \leq \sum_{i=1}^{k}\lambda_i(\varepsilon)$$

- In particular, If $\lambda < r \leq \lambda_{max}$ we have a "self scattering".

Figure 3.26. *Scattering choice*

4/ selected scattering effect

⌘ The interactions redistribute:

- wave vector direction
- velocity.

⌘ Re-direction of the wave vector is determined at random.

⌘ Using probability p(ε,θ) so that an energy particle will be deflected at an angle 0, this probability depending on the type of interaction.

⌘ So that an energy particle will be deflected at an angle 0, this probability depending on the type of interaction.

Figure 3.27. *Selected scattering effect*

5/ Screening

⌘ **In a strong accumilation**

⌘ **For a strobllz,**

^these interactions are for instance by colubia,d – cf. Brooks-Herring,)

⌘ **Limited by the Pauli exclusion which allows us to take into account the final state of a particle**

Figure 3.28. *Screening between particle and ionized atoms*

CONSERVATIONS LAWS DERIVED FROM BOLTZMANN EQUATIONS (recalls)

(1) Concentrations

$$\int_{R^3}\left(\frac{\partial f_v}{\partial t}+V_v(k).\frac{\partial f_v}{\partial x}-\frac{q}{\hbar}E(x,t).\frac{\partial f_v}{\partial k}\right)dk=\int_{R^3}\frac{f_v-f_{v0}}{\tau_v}dk$$

(2) Velocities

$$\int_{R^3}V_v(k)\left(\frac{\partial f_v}{\partial t}+V_v(k).\frac{\partial f_v}{\partial x}-\frac{q}{\hbar}E(x,t).\frac{\partial f_v}{\partial k}\right)dk=\int_{R^3}V_v(k)\left(\frac{f_v-f_{v0}}{\tau_v}\right)dk$$

(3) Energies

$$\int_{R^3}\varepsilon_v(k)\left(\frac{\partial f_v}{\partial t}+V_v(k).\frac{\partial f_v}{\partial x}-\frac{q}{\hbar}E(x,t).\frac{\partial f_v}{\partial k}\right)dk=\int_{R^3}\varepsilon_v(k)\left(\frac{f_v-f_{v0}}{\tau_v}\right)dk$$

Figure 3.29. *Conservations laws*

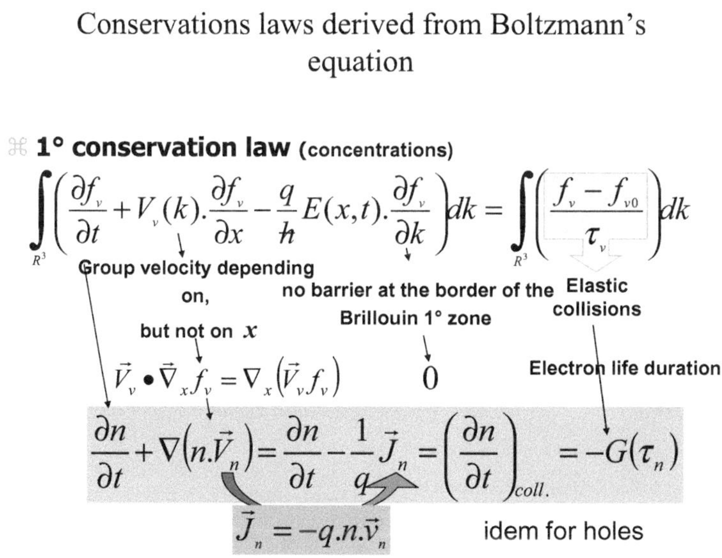

Figure 3.30. *Conservations laws. For a color version of this figure, see www.iste.co.uk/gontrand/nanoelectronics.zip*

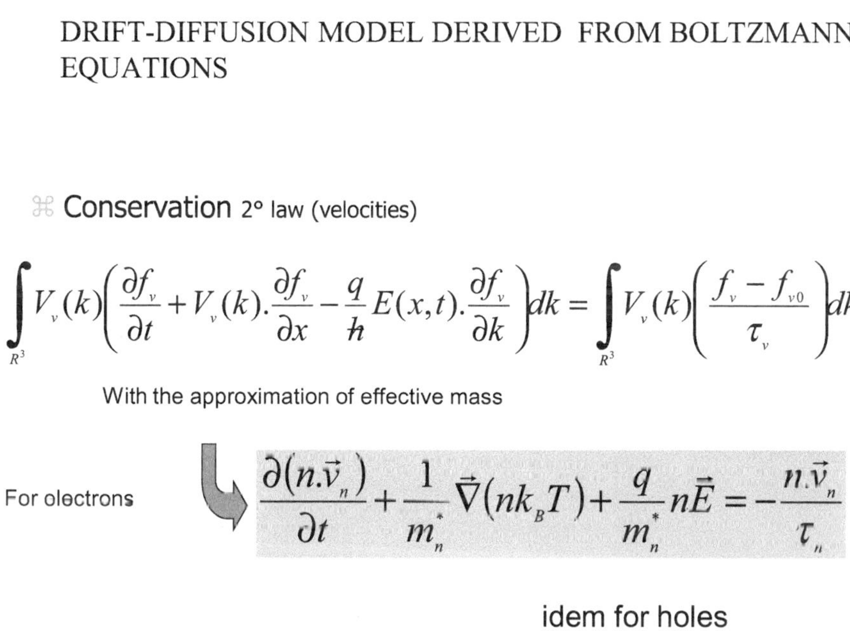

Figure 3.31. *From Boltzmann equation to drift-diffusion model*

2° law

⌘ Mobilities $\mu_n = \frac{q\tau_n}{m_n^*}$ $\mu_p = \frac{q\tau_p}{m_p^*}$

⌘ Current densities $\vec{J}_n = -q.n.\vec{v}_n$ $\vec{J}_p = +q.p.\vec{v}_p$

⌘ Uniform temperature

⌘ (carriers and crystalline network)

For electrons

$$\frac{\partial(n.\vec{v}_n)}{\partial t} + \frac{1}{m_n^*}\vec{\nabla}(nk_BT) + \frac{q}{m_n^*}n\vec{E} = -\frac{n.\vec{v}_n}{\tau_n}$$

$$\tau_n \frac{\partial \vec{J}_n}{\partial t} + \vec{J}_n - q\mu_n\left(\frac{k_BT}{q}\vec{\nabla}n + n.\vec{E}\right) = 0$$

idem for holes

Figure 3.32. *From Boltzmann equation to drift-diffusion model including temperature*

DRIFT-DIFFUSION MODEL DERIVED FROM BOLTZMANN EQUATIONS

⌘ Development versus the little parameter τ_n

$$\vec{J}_n = \sum_{i=0}^{\infty} \vec{J}_{ni}\tau_n^i$$

At 0 order

$$\vec{J}_n = q\mu_n\left(n.\vec{E} + \frac{k_BT}{q}\vec{\nabla}n\right)$$

$$\vec{J}_p = q\mu_p\left(p.\vec{E} - \frac{k_BT}{q}\vec{\nabla}p\right)$$

⇒ Drift diffusion with (Einstein relation) $D_{n,p} = \mu_{n,p}\frac{k_BT}{q}$

Figure 3.33. *From Boltzmann equation to generalized drift-diffusion model*

HYDRODYNAMIC MODEL

⌘ Compared to a conventional drift-diffusion model, we use 3 conservation laws in which carrier energy and temperature are unknown additional values.

⌘ To close the system, we link energy to carrier temperature using relations (in case of parabolic bands):

$$\xi_n = \frac{1}{2}m_n^*v_n^2 + \frac{3}{2}k_BT_n \qquad \xi_p = \frac{1}{2}m_p^*v_p^2 + \frac{3}{2}k_BT_p$$

⌘ Thus, carrier temperatures are no longer confused with network temperatures T_L and become unknown elements of the problem.

Figure 3.34. *Hydrodynamic model: temperature introduction*

HYDRODYNAMIC MODEL

⌘ We assume that the term for collision on energy can be expressed in the same manner as the term for collision on velocity.

$$\left(\frac{\partial \xi}{\partial t}\right)_{coll.} = -\frac{\xi - \xi_0}{\tau_e(\xi)} \qquad \tau_e = \text{relaxation time of energy.}$$

for electrons:

$$\frac{\partial n}{\partial t} + \nabla(nv_n) = \left(\frac{\partial n}{\partial t}\right)_{coll.} = -G(\tau_n)$$

$$\frac{\partial v_n}{\partial t} + v_n \nabla v_n + \frac{qE}{m_n^*} + \frac{1}{n.m_n^*}\nabla(n.k_B T_n) = \left(\frac{\partial v_n}{\partial t}\right)_{coll.} = -\frac{v_n}{\tau_n(\xi_n)}$$

$$\frac{\partial \xi_n}{\partial t} + v_n \vec{\nabla}\xi_n + qvE + \frac{1}{n}\nabla(n.v_n.k_B T_n) + \frac{1}{n}\nabla Q = \left(\frac{\partial \xi_n}{\partial t}\right)_{coll.} = -\frac{\xi_n - \xi_{n0}}{\tau_e(\xi_n)}$$

idem for holes

Figure 3.35. *Hydrodynamic model: starting equations*

HYDRODYNAMIC MODEL

⌘ Heat flux Q is :

$$Q = -\kappa \vec{\nabla} T_n$$

κ = thermal conductivity of the carriers .

$$\frac{\partial \xi}{\partial t} + qv\left(\frac{m^* v}{q\tau_n} + \frac{k_B T_n}{q}\frac{\vec{\nabla} n}{n} - \vec{\nabla}\left(\frac{3}{2}\frac{k_B T_n}{q}\right)\right) = -\frac{\xi - \xi_0}{\tau_e(\xi)}$$

for electrons

Figure 3.36. *Hydrodynamic model: with heat flux*

HYDRODYNAMIC MODEL

⌘ We can also reformulate the problem, by changing the velocity by the current density vector:

$$J_n = -q.n.v_n$$

⌘ Introducing the mobility as in the drift –diffusion model

⌘ The heat flux can be expressed as:

$$\frac{\partial J_n}{\partial t} = (J_n \vec{\nabla} + \nabla . J_n) J_n + \frac{1}{m}\nabla(nk_B T) + \frac{q}{m} nE = \left(\frac{\partial J_n}{\partial t}\right)_{coll.}$$

$$\overrightarrow{s_n} = -\frac{5r_n}{2}\left(\frac{k_B T_n}{q}\overrightarrow{J_n} + f_n^{hf}\kappa_n \vec{\nabla} T_n\right) \qquad \kappa_n = k_B^2 n \mu_n T_n$$

$$\overrightarrow{s_n} = -\frac{5r_p}{2}\left(\frac{-k_B T_p}{q}\overrightarrow{J_p} + f_p^{hf}\kappa_p \vec{\nabla} T_p\right) \qquad \kappa_p = k_B^2 p \mu_p T_p$$

Figure 3.37. *Hydrodynamic model; final model*

HYDRODYNAMIC MODEL

- The current density expression should also be reviewed to take tempearture variationsinto account:

$$\overrightarrow{J_n} = -\mu_n\left(qn\overrightarrow{\nabla\Phi} - k_B T_n \vec{\nabla} n - f_n^{td} k_B n \vec{\nabla} T_n\right)$$
$$\overrightarrow{J_p} = -\mu_p\left(qp\overrightarrow{\nabla\Phi} + k_B T_p \vec{\nabla} p + f_p^{td} k_B p \vec{\nabla} T_p\right)$$

- Finally, if we do not overlook the Joule effect, an additional equation must be added with network temperature T_L as unknown:

$$\frac{\partial T_L}{\partial t} = \nabla k_{Ln} T_L \quad \text{with appropriate boundary conditions}$$

Figure 3.38. *Hydrodynamic model: current density*

3.2.2. *Rapid devices for radiofrequency or microwave systems*

Heterojunction bipolar transistors (HBTs) emerged in order to meet the high demand for telecommunications, broadband and mega or gigabytes. Devices with SiG(C) alloy are far more rapid than their only-silicon counterparts. In fact, SiGe HBT, having a very fine SiGe epitaxial base, is the most rapid device available to date over CMOS, with a cut-off frequency higher than 400 GHz. Moreover, a SiGe high-speed bipolar element is quite appealing for wireless and optic (numerical) telecommunications when integrated with advanced BiCMOS (Bipolar Compatible CMOS) technology.

Faced with bipolar transistor limitations for high-frequency applications, adding germanium to the base was the easiest solution in terms of performance/cost. The SiGe heterojunction improved the performance of silicon-based bipolar transistors and made them competitive with III/V groups for high-frequency applications [ORI 99]. The appeal of the $Si_{1-x}Ge_x$ alloy (IV–IV group), compared with the III–V material, lies in the fact that the microelectronic industry is mainly based on silicon; the technology is ready and the development cost is minimal.

The main appeal of the $Si_{1-x}Ge_x$ material is that it has an energy bandgap that is smaller than that of Si. This property is applied to bipolar transistor bases for differentiating the potential barriers and forces applied to electrons and holes. This will improve the efficiency of electron injection from the emitter into the base, all while blocking holes better (see NPN transistor). Injecting electrons from the emitter into the base is favored, allowing a higher collector current. The result is that SiGe HBTs have a larger current gain than "fully Si" ones.

3.2.2.1. *SiGeC HBT*

DDM modeling (drift diffusion model)

We can analyze the electrical behavior of the HBT towards semiconductors using fundamental equations. Our analysis will be based on a macroscopic description of semiconductors with a possible non-uniform composition. Semiconductors differ in fundamental properties such as energy bandgap, carrier mobility, effective mass of electrons and holes. Moreover, interfaces between different materials must be described correctly.

i) For modeling the SiGeC heterostructures, we use standard drift diffusion equations. Poisson's equation must show that the dielectric constant is a function of position. In continuity equations, generation and recombination rates must describe the intrinsic carrier concentration, which depends on the position. The most important modification to standard semiconductor equations is given in the Poisson equation:

$$\nabla^2 \phi = \frac{-q}{\varepsilon_{S/C}} \left[p - n + N_D^+ - N_A^- \right] \qquad [3.1]$$

ii) Continuity equations for electrons and holes:

$$\frac{\partial n}{\partial t} = GR_n + \frac{1}{q} \frac{dJ_n}{dx} \qquad [3.2]$$

$$\frac{\partial P}{\partial t} = GR_p - \frac{1}{q} \frac{dJ_p}{dx} \qquad [3.3]$$

Current density equations for electrons and holes:

$$J_n = -q n \mu_n \frac{d\phi_n}{dx} \qquad [3.4]$$

$$J_p = q p \mu_p \frac{d\phi_p}{dx} \qquad [3.5]$$

Here, N^+_D and N^-_A are ionized impurity concentrations, ε is the dielectric constant of the material and q is the absolute value of the charge related to an electron or a hole. Electron and hole current densities are functions of concentrations, μ_n and μ_p are the electron and hole carrier mobilities, and the quasi-Fermi potential for electrons and holes are Φ_n and Φ_p, respectively.

First, the simulator solves these partial differential equations (PDE) for the electrostatic potential Φ and for electron and hole concentrations n and p, respectively. GR_n and GR_p are the net recombination rates for holes or electrons, respectively. The recombination–generation models for Si/e SiGeC heterojunctions are the same as the models described previously for homojunctions; in this case, we use the well-known Shockley–Read–Hall (SRH) model:

$$GR_n = GR_p = \frac{n \cdot p - n_i^2}{\tau_n \left(p + p_i\right) + \tau_p \left(n + n_i\right)} \quad [3.6]$$

τ_n and τ_p are the electron and hole life times in the semiconductor.

$$n_i^2{}_{SiGeC} = \left(N_C \cdot N_V\right)_{SiGec} \cdot \exp\left(-\frac{E_{g\ SiGec}}{KT}\right) \quad [3.7]$$

For the numerical implementation of (1)–(5), in 2D, the associated variables are t, potential Φ, electron n and hole p concentrations; the Scharfetter–Gummel discretization and the Gauss–Seidel method are used in the numerical solution. E_{FN} and E_{FP} are quasi-Fermi levels, and:

$$\phi_n = -\frac{1}{q} E_{FN},\ \phi_p = -\frac{1}{q} E_{FP} \quad [3.8]$$

$$E_{FN} = E_c + KT \ln\left(\frac{n}{N_c}\right) + KT \ln \gamma_n \quad [3.9]$$

$$E_{FP} = E_V - KT \ln\left(\frac{p}{N_v}\right) + KT \ln \gamma_p \quad [3.10]$$

Rq: $\gamma_n = \gamma_p = 1$ for a Boltzmann [3.11]

E_C and E_V are valence and conduction bands, and n and p are electron and hole concentrations, respectively, and γ_n and γ_p are parameters relating to Fermi–Dirac statistics (used in the case of nitrogen):

$$E_C = -q\phi + \frac{E_g}{2} \quad [3.12]$$

$$E_V = -q\phi - \frac{E_g}{2} + \Delta E_v \qquad [3.13]$$

ΔE_V is the valance band discontinuity; we do not know how this bandgap is distributed in the valence band of SiGe and in the conduction band of SiGeC. Initially, our simulations were made by assuming that this gap variation occurs in the valence band.

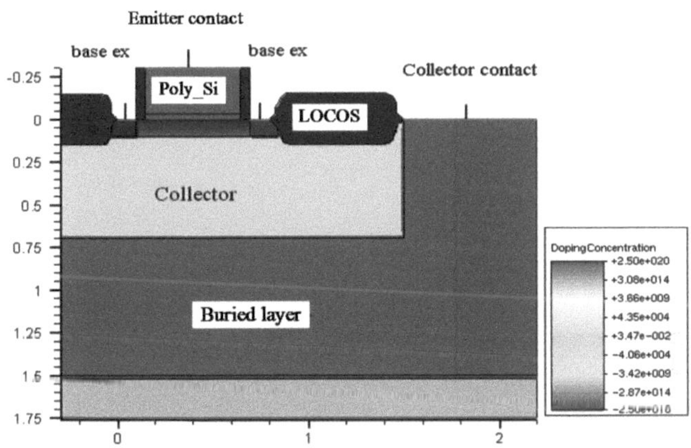

Figure 3.39. *Schematic of the transversal section of a SiGe (C) heterojunction bipolar transistor with 0.3 µm BiCMOS technology and polysilicon emitter and doping levels. For a color version of this figure, see www.iste.co.uk/gontrand/nanoelectronics.zip*

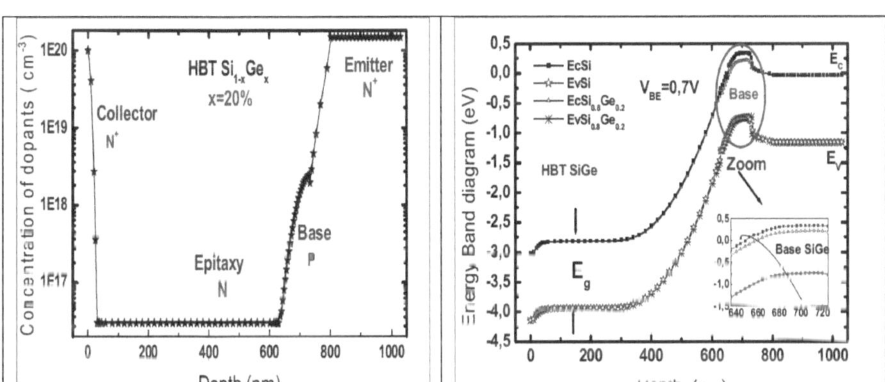

Figure 3.40. *Cross-section of a typical heterojunction bipolar transistor (HBT). For a color version of this figure, see www.iste.co.uk/gontrand/nanoelectronics.zip*

Electrical results

In the first step, we simulated the Gummel plot characteristics of a SiGe transistor (I_C, I_B as a function of V_{BE}) compared with the SiGeC equivalent (same geometry, same profiles). One of the objectives is to analyze the impact of carbon concentration on electrical characteristics of the device.

We studied the effect of a low carbon concentration (0.5, 0.75 and 1%) in a SiGe HBT device. An important outcome of adding carbon is the increase in the base current. This increase is due to the reduction in the minority carrier lifetime in the base, which is due to the incorporation of a little carbon in SiGeC and the presence of deep traps in the emitter–base junction. The drop in I_C can be explained by the increase in (band)gap, especially by the increase in boron dose in the neutral base region due to the decrease in boron diffusion.

Moreover, we study these transistors with or without defects induced by the carbon presence, at the neighboring of the E.B junction. Figure 3.41(a) shows the evolution of current gain, with or without these defects. The results obtained in this study are efficiently compared with the electrical characteristics obtained from SPICE simulations (see Kirchhoff equations) via the compact model HICUM (HIgh CUrrent Model) integrated in the standard ADS (Advanced Design System) simulator. "Gummel" and "Early" are shown in Figures 3.41(a) and (b), respectively. We present these simulated characteristics compared with those of the commercial simulator ADS ($0.17 \times 6.2\ \mu m^2$) for different transistors: $0.17 \times 1.9\ \mu m^2$ and $0.17 \times 19.9\ \mu m^2$.

Figure 3.41(b) illustrates I_C–V_{CE} characteristics (Early curves) for a SiGeC HBT with 0.75% carbon concentration, simulated with our algorithm and commercial simulator ADS (ref) ($0.17 \times 6.2\ \mu m^2$) as well as compared with transistor measurements ($0.17 \times 5.9\ \mu m^2$).

Electrical performance of this device can be affected by the presence of inherent defects in complex structure shrinking. In our devices, most of these defects are in the vertical interface between the spacers and the polysilicon emitters, arising from the RIE (Reactive Ion Etching) process. Their location, as well as their density or capture cross-section, has an influence on the electrical characteristics of the HBT. Figure 3.41 shows the Gummel characteristics for two similar SiGeC HBTs: one with hole traps at 0.6 eV (reminder: mi-gap = 0.56 eV), density $N_T = 10^{18}$ cm^{-3} and capture cross-section $\sigma = 10^{17}$ cm^2 – according to the DLTS (Deep-Level Transient Spectroscopy) measurements – and another considered flawless.

These defects cause noise phenomena, especially in nanoelectronics, which is covered in Chapter 4.

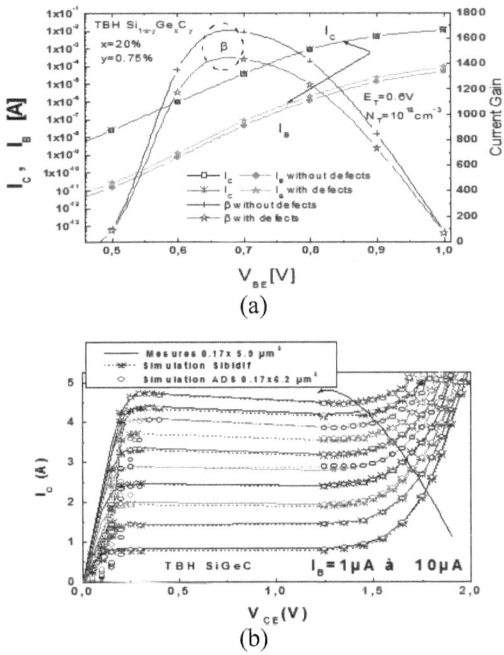

Figure 3.41. *a) Gummel characteristics: I_C, I_B and current gain as a function of V_{BE} for the HBT, compared with measurements. b) Early characteristics: IC as a function of V_{CE} for a SiGeC HBT, compared with the commercial simulator ADS (0.17×6.2 µm^2) and transistor measurements (0.17×5.9 µm^2) [LAK 08]. For a color version of this figure, see www.iste.co.uk/gontrand/nanoelectronics.zip*

Hereafter, we present dynamic electrical parameters such as the cut-off frequencies of the HBT.

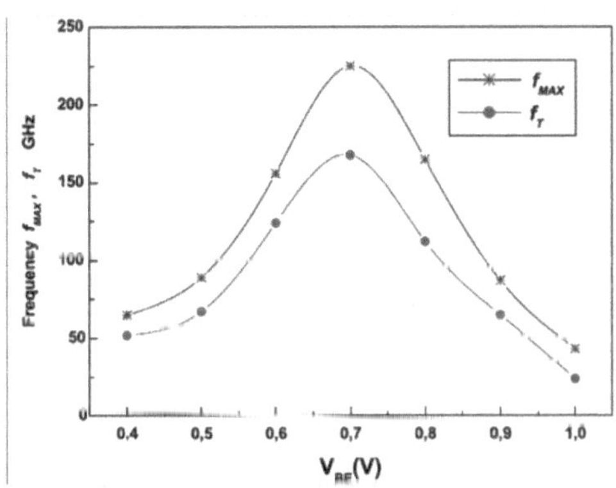

Figure 3.42. *Cut-off frequencies of the HBT*

3.2.3. *DGMOSFET*

3.2.3.1. *Introduction*

For MOS transistors, parasitic effects can degrade electrical performances. For example, source-to-channel coupling [CLE 01] can be degraded by transverse quantum confinement; this corresponds to a tunnel-effect passage through the gate, leading to an increase in power consumption [BOU 13]. On the contrary, source–drain channel coupling can be considered a critical physical limitation to effective MOS performance. For this reason, we consider integrating new devices that function according to quantum laws rather than classical physical laws. In this context, RTDs (Resonant Tunneling Diodes) have pertinent characteristics.

Nevertheless, the immediate future of microelectronics might have delays arising from the MOS, CMOS (see ultimate CMOS) or BiCMOS device; it depends on the industry's capacity and that of researchers to pursue integration and miniaturization. In the case of continuous miniaturization, it would be interesting to look at new transistor architectures [THO 09].

DGMOS (Double-Gate MOS) might seem to be the best choice for nano-oscillators for the industry; compatible with the best current MOS technologies, it allows for impedance matching and limits inconvenient thermal constraints.

The introduction of new gates, with respect to conventional MOS transistors, is useful for device performance in several ways. The advantage of the SOI (Silicon On Insulator) is added to other performance enhancements thanks to multiple gates. Among the various advantages, a notable one is the enhanced threshold characteristics of transistors (increase in the I_{ON}/I_{OFF} ratio, increase in the drain current or even better carrier mobility in the channel).

Despite these advantages, technological progress is necessary for optimizing these devices. In particular, industrial manufacturing of double-gate transistors still remains a challenge. The main interest in a double-gate MOSFET is efficiently controlling the silicon channel with a very small channel width. This means including a gate contact on either side of the channel.

However, even double-gate transistors have limitations when the dimensions are at the nano level. This miniaturization has a direct repercussion on silicon. Silicon dioxide (S_iO_2) reaches its dimension limitations; thus, new architectures must be found.

In this section, we present a flexible and efficient numerical DGMOS model based on a quantum description of carrier concentration. It provides a self-consistent solution to Schrödinger and Poisson equations. Physical phenomena such as carrier confinement or quasi-ballistic transport are taken into account. The double gate is used exactly for creating an integrated DGMOS (nanometric channel and ultra-thin silicon films). The orientation of silicon is <100>; the DGMOS has a symmetric gate and an n-type channel.

Calculation of the electrical field and the carrier concentration was performed simply using a finite difference scheme with a uniform grid; coupled equations are block-solved via the Newton–Raphson method.

3.2.3.2. *Structure*

Figure 3.43 illustrates a DGMOSFET.

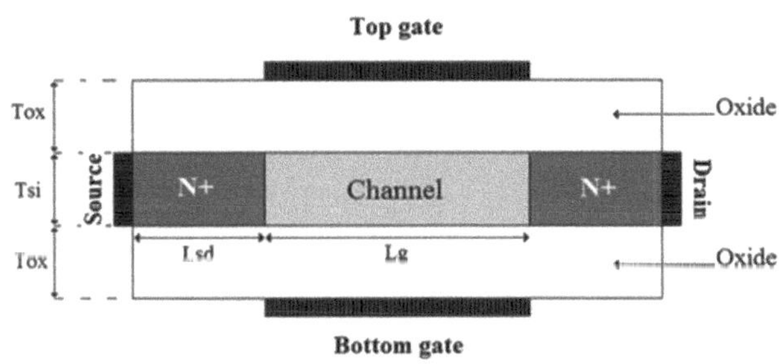

Figure 3.43. *DGMOS transistor diagram*

Electron transport takes place inside the channel, which is the active area outside the source and drain reservoirs. We have taken a DGMOSFET with the following parameters: doping $N_A = 10^{10}$ cm^{-3}, source/drain doping $N+ = 10^{20}$ cm^{-3}, silicon thickness TSi = 1.5 nm, oxide thickness TOX = 1.5 nm, channel length LG = 10 nm and source/drain length L_{SD} = 5 nm. Aluminum was used for metallization, whose work function was 4.35 eV. Identical V_{GS} voltages were applied to both gates. All calculations were verified at ambient temperature (T = 300 K).

Nano-level transport in MOSFETs

Electrical transport in a material is random (stochastic). Electrons (classically described) undergo collisions (with defects, phonons and other electrons). Moreover, when the MOS transistor reaches its nanometric dimensions, carrier transport inside the active area (channel) changes, as the channel length is around the average distance between the two interactions, called the mean free path. A large number of carriers are thus able to pass through the channel without undergoing collision. Electrons can move without diffusion as in an empty tube; in this case, the transport is ballistic [CER 88].

Electron density

The energy profile in a ballistic MOS transistor is illustrated in Figure 3.44.

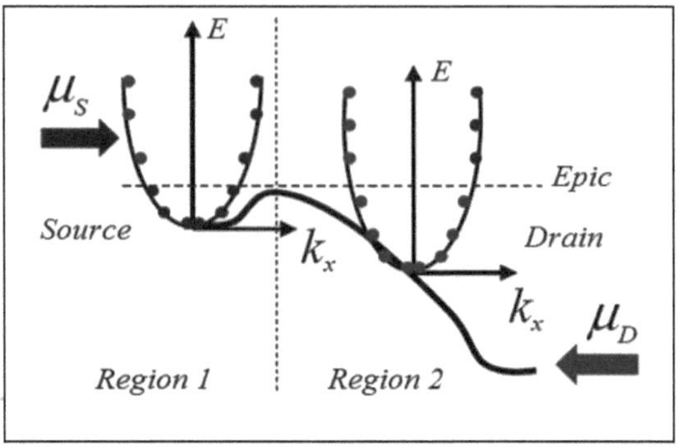

Figure 3.44. *Evolution of the conduction band along the source–drain axis in ballistic transport*

This profile can be divided into two areas: peak energy points on the left (area 1) and Epic peak energy points on the right (area 2). In area 1, electrons with energy lower than Epic come from the drain reservoir. This explanation can be applied to area 2. In fact, each carrier is linked to an incident wave and a reflected wave (see quantum mechanics). Both reservoirs (source, drain) are characterized by two Fermi levels: μ_S and μ_D. In area 1, electron density can be written as:

$$\begin{aligned} n_{left}(x, E_{kj}) = & \int_0^{\infty} \left[\frac{1}{\pi\bar{h}} \sqrt{\frac{m_x^*}{2E_x}} \frac{1}{1 + \exp((E_x + E_i + E_{kj} - \mu_s)/k_B T)} \right] dE_x \\ & + \int_0^{E_{peak}} \left[\frac{1}{\pi\bar{h}} \sqrt{\frac{m_x^*}{2E_x}} \frac{1}{1 + \exp((E_x + E_i + E_{kj} - \mu_s)/k_B T)} \right] dE_x \\ & + \int_{E_{peak}}^{\infty} \left[\frac{1}{\pi\bar{h}} \sqrt{\frac{m_x^*}{2E_x}} \frac{1}{1 + \exp((E_x + E_i + E_{kj} - \mu_D)/k_B T)} \right] dE_x \end{aligned} \quad [3.14]$$

The left index is for area 1.

Taking into account the transverse mode contributions to the integration over Ek_j, we get:

$$n_{left}(x) = \int_0^{\infty} \frac{1}{\pi\bar{h}} \sqrt{\frac{m_y^*}{2E_{kj}}} \left[n_{left}(x, E_{kj}) \right] dE_{kj} \quad [3.15]$$

$$n_{left}(x) = n_{2Di}\left\{\ln(1+e^{\bar{\mu}_s}) + \frac{1}{\sqrt{\pi}}\int_0^{\tilde{E}_{peak}}\frac{d\tilde{E}_x}{\sqrt{\tilde{E}_x}}\Im_{-\frac{1}{2}}\left(\tilde{\mu}_s-\tilde{E}_x\right) + \frac{1}{\sqrt{\pi}}\int_{\tilde{E}_{peak}}^{\infty}\frac{d\tilde{E}_x}{\sqrt{\tilde{E}_x}}\Im_{-1/2}\left(\tilde{\mu}_D-\tilde{E}_x\right)\right\} \quad [3.16]$$

where ~ means that all quantities are expressed with respect to the local sub-band Ei (x) and standardized with respect to the thermal energy k_{BT} and n_{2Di}: the electron density surface for each sub-band is found using [MOG 90]:

$$n_{2Di} = \frac{\sqrt{m_x^* m_y^*}}{\pi\bar{h}^2}\frac{k_B T}{2} \quad [3.17]$$

Similarly, electron density in area 2 can be expressed as:

$$n_{right}(x) = n_{2Di}\left\{\ln(1+e^{\bar{\mu}_D}) + \frac{1}{\sqrt{\pi}}\int_0^{\tilde{E}_{peak}}\frac{d\tilde{E}_x}{\sqrt{\tilde{E}_x}}\Im_{-1/2}\left(\tilde{\mu}_D-\tilde{E}_x\right) + \frac{1}{\sqrt{\pi}}\int_{\tilde{E}_{peak}}^{\infty}\frac{d\tilde{E}_x}{\sqrt{\tilde{E}_x}}\Im_{-1/2}\left(\tilde{\mu}_s-\tilde{E}_x\right)\right\} \quad [3.18]$$

and the ballistic current density is expressed as:

$$\begin{aligned}J(E_{kj}) = {} & \int_0^{E_{peak}}\left[\sqrt{\frac{2E_x}{m_x^*}}\frac{q}{\pi\bar{h}}\sqrt{\frac{m_x^*}{2E_x}}\frac{1}{1+\exp\left(\frac{E_x+E_i+E_{kj}-\mu_s}{k_B T}\right)}\right]dE_x + \\ & \int_{E_{peak}}^{\infty}\left[\sqrt{\frac{2E_x}{m_x^*}}\frac{q}{\pi\bar{h}}\sqrt{\frac{m_x^*}{2E_x}}\frac{1}{1+\exp\left(\frac{E_x+E_i+E_{kj}-\mu_s}{k_B T}\right)}\right]dE_x - \\ & \int_0^{E_{peak}}\left[\sqrt{\frac{2E_x}{m_x^*}}\frac{q}{\pi\bar{h}}\sqrt{\frac{m_x^*}{2E_x}}\frac{1}{1+\exp\left(\frac{E_x+E_i+E_{kj}-\mu_s}{k_B T}\right)}\right]dE_x - \\ & \int_{E_{peak}}^{\infty}\left[\sqrt{\frac{2E_x}{m_x^*}}\frac{q}{\pi\bar{h}}\sqrt{\frac{m_x^*}{2E_x}}\frac{1}{1+\exp((E_x+E_i+E_{kj}-\mu_D)/k_B T)}\right]dE_x\end{aligned} \quad [3.19]$$

where $\sqrt{2E_x/m_x^*}$ represents the drift velocity and q represents the elementary charge.

As there are opposite signs for electron velocity, equation [3.6] is simplified as.

$$J(E_{kj}) = \frac{q}{\pi\hbar}\int_{E_{peak}}^{\infty}\left[\frac{1}{1+\exp((E_x+E_i+E_{kj}-\mu_S)/k_BT)} - \frac{1}{1+\exp((E_x+E_i+E_{kj}-\mu_D)/k_BT)}\right]dE_x \quad [3.20]$$

Integrating over E_{kj}, we get:

$$J = \int_0^{\infty}\frac{1}{\pi\hbar}\sqrt{\frac{m_y^*}{2E_{kj}}}J(E_{kj})dE_{kj} \quad [3.21]$$

$$J = \frac{q}{\hbar^2}\sqrt{\frac{m_y^*}{2}}\left(\frac{-k_BT}{\pi}\right)^{3/2}\left[\Im_{1/2}(\tilde{\mu}_s - \tilde{E}_{pic}) - \Im_{1/2}(\tilde{\mu}_D - \tilde{E}_{peak})\right] \quad [3.22]$$

The total density is the sum of all contributions of the valleys and sub-bands.

Numerical methods

Reducing the size of devices until they reach nanometric dimensions reveals the nature of quantum phenomena. The devices affect transport significantly, and thus the simulator must take this into account efficiently. To determine the electrical characteristics of this device, a mathematical model that describes the transport of electronic support, taking into account existing quantum phenomena, must be established.

There are various formalisms, of which we will present what seems the most appropriate to us.

3.2.3.3. *Wigner function*

At the nano level, the wave nature of electrons comes into play and the semi-classical Boltzmann approach to transport, usually used for modeling device behavior, is insufficient. Extending this formalism to the quantum level mainly leads to the use of the Wigner transport equation [SIL], which significantly resembles the Boltzmann equation.

Using the "Wigner" equation is then one of the possible formalisms dedicated to quantum formalism. Other studies have been realized [SHI 01, ORI 99].

The Wigner equation is derived from the Schrödinger equation and the density matrix functional (DMF). This involves combining the Schrödinger equation and DMF, and then performing the Fourier transform [ROS 98a]. In fact, the Schrödinger equation defines the wave function of an electron moving in a given potential profile with a certain energy. DMF is an operator that links the wave function in different points in real space and "weighs" using a Fermi–Dirac distribution, through the wave vector; the solution of the Wigner function corresponds to a wave function distribution through phase space.

3.2.3.4. *Formalism*

A single mass particle m is considered to move under the potential V(x). The phase space is defined by the Cartesian product of position x and moment p. Physical quantities are dynamic functions A(x, p) of phase space coordinates, such as kinetic and potential energies and their sum from which we get Hamiltonian H(x, p). The state of the particle at a given moment is represented by a point in the phase space. Provided that the initial coordinates of particles are known, coordinates x(t) and p(t), for a particular t time, are found using Hamilton equations:

$$\dot{x} = \frac{\partial H(x,p)}{\partial p} = \frac{p}{m} \qquad [3.23]$$

$$\dot{p} = \frac{\partial H(x,p)}{\partial x} = -\frac{\partial V(x)}{\partial x} \qquad [3.24]$$

Function A(t) describes how physical quantities vary according to time. There are two possibilities: (a) A(t) = A(x (t), p(t)) is the old function of new coordinates and (b) A(t) = A(t, x, p) is the new function of old coordinates. In the first case, we postulate that laws of mechanics do not change with time: A remains the same for old and new coordinates. Then, using equations [3.10] and [3.11], we get the evolution equation of A:

$$\dot{A} = \frac{\partial A(x,p)}{\partial x}\frac{\partial H(x,p)}{\partial p} - \frac{\partial A(x,p)}{\partial p}\frac{\partial H(x,p)}{\partial x} = [A, H]_P, [x, p]_P = 1 \qquad [3.25]$$

The Poisson bracket [•, •] P is a fundamental notion in dynamic functions. It gives rise to automorphic mapping (preserving the algebraic structure) of the whole set of functions.

On the one hand, in the second case, we must postulate a law for the evolution of A(t, x, p). If it is applied according to equation [3.12], automorphism systematically conserves laws of mechanics: the new function of old coordinates is the old function of new coordinates: A(t, x, p) = A(x (t), p(t)).

Statistic description is introduced if the coordinates of a point cannot be defined exactly but with a certain probability. According to the basic premise of classical statistical mechanics, the state of a particle system is fully defined by function f(x, p), with the following properties:

$$f(x,p) \geq 0; \int dxdpf(x,p) = 1 \qquad [3.26]$$

Physical quantities are defined using the corresponding average values:

$$\langle A \rangle(t) = \int dxdpA(t,x,p)f(x,p) \qquad [3.27]$$

This equation is not practical as it requires calculating the evolution of a particular quantity A. However, due to automorphism of the Poisson bracket, it is possible to modify the variables such that time is transformed into distribution function f.

Equation [3.14] becomes:

$$\langle A \rangle(t) = \int dxdpA(x,p)f(x,p,t) \qquad [3.28]$$

The evolution equation for f can be derived using [3.10], [3.11] and [3.12]:

$$\left(\frac{\partial}{\partial t} + \frac{p}{m}.\frac{\partial}{\partial x} + F(x)\frac{\partial}{\partial p}\right) f(x,p,t) = \left(\frac{\partial f}{\partial t}\right)_C \qquad [3.29]$$

Here, force F = −∇xV is obtained using the spatial derivative of the potential energy. The characteristics of the differential operator in brackets, called the Liouville operator, are classical Newton trajectories obtained from equations [3.10] and [3.11]. During these trajectories, the left-hand side of equation [3.16] becomes a total derivative with respect to time: in the case of zero interaction with the environment (∂f/∂t)t = 0, note that the trajectories do not change. Otherwise, particles are redistributed between trajectories, and the right-hand side of equation [3.16] is equal to the net change of particle density due to collisions. In the rest of this section, we get a quantum analogue of [3.13], [3.15] and the Boltzmann equation [3.16].

Quantum operators

We recall the principles of operators in quantum mechanics, which will be used for the phase space formalism. Hermitian operators Â represent the physical quantities in quantum mechanics:

$$\hat{A}|\phi_n\rangle = a_n|\phi_n\rangle; \ \langle \phi_n|\phi_m\rangle = \delta_{mn} \qquad \sum_n |\phi_n\rangle\langle\phi_n| = \hat{1} \qquad [3.30]$$

These operators have actual values and a complete system of orthogonal vectors that form the abstract Hilbert space; the states of the system are specified by elements $|\Phi_t\rangle$ of the Hilbert space H that are square-integrable functions and not standardized with respect to norm L_2 in H. In wave mechanics, it is postulated that the evolution of | ψ_t> is provided by the Schrödinger equation:

$$\hat{H}\,|\,\Psi_t\rangle = i\hbar\frac{\partial\,|\,\Psi_t\rangle}{\partial t}\langle\Psi_t\,|\,\Psi_t\rangle = 1\,|\,\Psi_t\rangle = \sum_n c_n(t)\,|\,\phi_n\rangle \qquad [3.31]$$

The state might be decomposed into a full basis of an observable A. Moreover, it can be demonstrated that during evolution, the state is standardized. This property is often called probability conservation. According to the correspondence principle, Hermitian operators x and p correspond to the classical position and impulsion variables, satisfying quantum counterparts of the Poisson bracket:

$$x \to \hat{x} \qquad p \to \hat{p} \qquad \hat{x}\hat{p} - \hat{p}\hat{x} = [\hat{x}, \hat{p}]_- = i\hbar\hat{1} \qquad [3.32]$$

Wave mechanics only uses half of the phase space – coordinates or moment – to describe a physical system.

According to [3.17] and [3.19], the coordinates are:

$$\hat{x}|x\rangle = x|x\rangle \quad \int dx|x\rangle\langle x| = \hat{1} \quad \hat{p} = -i\hbar\frac{\partial}{\partial x} \qquad [3.33]$$

Finally, we recall the equation for the average value of a physical quantity:

$$\langle A\rangle(t) = \langle \Psi_t|\hat{A}|\Psi_t\rangle = \int dx \langle \Psi_t|x\rangle\langle x|\hat{A}|\Psi_t\rangle \qquad [3.34]$$

Operation formulation of quantum mechanics seems too abstract when compared with known classical concepts. Nevertheless, it is possible to reformulate quantum mechanics ideas in phase space. The first step involves evaluating the actual number of variables involved in [3.21]. Using [3.17] and [3.20], the first step is:

$$\langle x|\hat{A}|\Psi_t\rangle = \int dx' \sum_n a_n \langle x|\phi_n\rangle\langle\phi_n|x'\rangle\langle x'|\Psi_t\rangle = \int dx' \alpha(x,x')\Psi_t(x') \qquad [3.35]$$

where $\psi_t(x) = \langle x|\psi_t\rangle$. A substitution in [3.21] shows that the physical average has been effectively assessed in a "double half" of the phase space:

$$\langle A\rangle(t) = \int dx' \int dx \alpha(x,x')\rho_t(x',x) = Tr(\hat{\rho}_t \hat{A}) \qquad [3.36]$$

with ρ_t and $\hat{\rho}_t$, the density matrix and the density operator, respectively, given by:

$$\begin{aligned} \rho_t(x,x') &= \Psi_t^*(x')\Psi_t(x) = \langle x|\Psi_t\rangle\langle\Psi_t|x'\rangle = \langle x|\hat{\rho}_t|x'\rangle \\ \rho_t &= \sum_{m,n} c_m^*(t)c_n(t)|\phi_n\rangle\langle\phi_m| \end{aligned} \qquad [3.37]$$

3.3. Wigner function

Equation [3.18] and its associated equation lead to the von Neumann equation of motion for density matrix of the pure state ρ_t [3.24]:

$$i\hbar\frac{\partial\rho(x,x',t)}{\partial t} = \langle x|[\hat{H},\hat{\rho}_t] - |x'\rangle$$
$$= \left\{-\frac{\hbar^2}{2m}\left(\frac{\partial^2}{\partial x^2} - \frac{\partial^2}{\partial x'^2}\right) + (V(x) - V(x'))\right\}\rho(x,x',t) \quad [3.38]$$

Variables are modified using a center of mass change:

$$x_1 = \frac{(x+x')}{2};\ x_2 = x - x'$$

$$\frac{\partial\rho\left(x_1+\frac{x_2}{2},x_1-\frac{x_2}{2},t\right)}{\partial t}$$
$$= \frac{1}{i\hbar}\left\{-\frac{\hbar^2}{m}\frac{\partial^2}{\partial x_1\partial x_2} + (V(x_1+x_2/2) - V(x_1-x_2/2))\right\}\rho\left(x_1+\frac{x_2}{2},x_1\right.$$
$$\left.-\frac{x_2}{2},t\right) \quad [3.39]$$

The Wigner function is obtained using the Fourier transform with respect to x_2:

$$f_w(x_1,p,t) = \frac{1}{(2\pi\hbar)}\int dx_2\rho(x_1+x_2/2,x_1-x_2/2,t)e^{-ix_2\cdot p/\hbar} \quad [3.40]$$

We note that x_1 and p are independent variables due to the Wigner transform. It is easy to show that the corresponding operators commute. Thus, p x_1 and p define a Wigner phase space. The Fourier transform on the right-hand side of [3.26] leads to two terms that can be assessed as follows; it is convenient to use the following abbreviation:

$\rho(+,-,t)$ for $\rho\left(x_1+\frac{x_2}{2},x_1-\frac{x_2}{2},t\right)$:

$$I = -\frac{1}{i\hbar}\frac{\hbar^2}{m(2\pi\hbar)}\int dx_2 e^{-ix_2\cdot\frac{p}{\hbar}}\frac{\partial^2\rho(+,-,t)}{\partial x_1\partial x_2} = -\frac{1}{m}p.\frac{\partial f_w(x_1,p,t)}{\partial x_1} \quad [3.41]$$

where we integrated by parts and made use of the fact that the density matrix tends towards zero to infinity: $\rho \to 0$ if $x_2 \to \pm\infty$ 0

$$II = \frac{1}{i\hbar(2\pi\hbar)}\int dx_2 e^{-ix_2\cdot\frac{p}{\hbar}}(V(x_1+x_2/2) - V(x_1-x_2/2))\rho(+,-,t)$$
$$= \frac{1}{i\hbar(2\pi\hbar)}\int dx'e^{-ix_2\cdot\frac{p}{\hbar}\left(V\left(x_1+\frac{x_2}{2}\right)-V\left(x_1-\frac{x_2}{2}\right)\right)}.\delta(x_2-x')\rho\left(x_1+\frac{x'}{2},x_1\right. \quad [3.42]$$
$$\left.-x'/2\right)$$

After replacing the following integral with the delta function:

$$\delta(x_2 - x') = \frac{1}{(2\pi\hbar)} \int dp' e^{i(x_2 - x')p'/\hbar} \quad [3.43]$$

we get:

$$II = \frac{1}{i\hbar(2\pi\hbar)} \int dp' \int dx_2 e^{-ix_2 \cdot \frac{p-p'}{\hbar}} \left(V(x_1 + x_2/2) - V(x_1 - x_2/2)\right) \cdot \frac{1}{(2\pi\hbar)} \int dx' e^{-ix' \cdot \frac{p'}{\hbar}} \rho\left(x_1 + \frac{x'}{2}, x_1 - \frac{x'}{2}; t\right) = \int dp' V_w(x_1, p - p') f_w(x_1, p') t \quad [3.44]$$

We summarize the results of these transformations. Equation [3.26] leads to the Wigner equation:

$$\frac{\partial f_w(x,p,t)}{\partial t} + \frac{p}{m} \cdot \frac{\partial f_w(x,p,t)}{\partial x} = \int dp' V_w(x, p - p') f_w(x, p', t) \quad [3.45]$$

where V_w is the Wigner potential:

$$V_w(x,p) = \frac{1}{i\hbar(2\pi\hbar)} \int dx' e^{-ix' \cdot \frac{p}{\hbar}} \left(V(x + x'/2) - V(x - x'/2)\right) \quad [3.46]$$

A change in the sign of x' shows the antisymmetry of the Wigner potential.

Classical limitation of Wigner equation

We discuss the classical limitation of [3.46] by considering the case when potential V is a linear or quadratic function of position, where:

$$V\left(x \pm \frac{x'}{2}\right) = V(x) \pm \frac{\partial V(x)}{\partial x} \frac{x'}{2} + \cdots = V(x) \mp F(x) \frac{x'}{2} + \cdots \quad [3.47]$$

The dots represent the quadratic term. Force F can also be a linear function of position. As the even terms in the Taylor series expansion of V cancel out in [3.24], the Wigner potential becomes:

$$V_w(x,p) = \frac{1}{\hbar(2\pi\hbar)} \int dx' e^{-ix' \cdot \frac{p}{\hbar}} F(x) x' \quad [3.48]$$

The right-hand side of [3.32] becomes:

$$\int dp'\, V_w(x, p-p') f_w(x,p',t) = \\ \frac{i}{\hbar(2\pi\hbar)} \int dp' \int dx' e^{-ix'\frac{p-p'}{\hbar}} F(x) x' f_w(x,p',t) = \\ \frac{-F(x)}{2\pi\hbar}\frac{\partial}{\partial p} \int dp' \int dx'\, e^{-ix'\frac{p-p'}{\hbar}} f_w(x,p',t) = -F(x)\frac{f_w(x,p,t)}{\partial p} \quad [3.49]$$

having taken into account:

$$ix' e^{-ix'.\frac{p-p'}{\hbar}} = -\hbar\frac{\partial}{\partial p} e^{-ix'\frac{p-p'}{\hbar}} \quad [3.50]$$

Thus, the Wigner equation is reduced to a Boltzmann equation without the term for collision:

$$\frac{\partial f_w(x,p,t)}{\partial t} + \frac{p}{m}.\frac{\partial f_w(x,p,t)}{\partial x} + F(x)\frac{\partial f_w(x,p,t)}{\partial p} = 0 \quad [3.51]$$

Let us now consider, as an initial condition, a minimum uncertainty on wave packets. The Wigner function of such a packet is a Gaussian of the position and the moment. This can also be interpreted as an initial classical distribution of electrons. Provided that the force is a constant or linear function of position, the packet evolves according to [3.37]. The evolution resembles that of classical distribution. Despite phase space expansion, Gaussian devices determine the general form of the packet. During the evolution, f_w remains positive. However, strong changes to the electric field depending on the position can lead to interference effects. Beside the band offsets, the packet quickly loses its initial form and negative values appear.

Numerical approach: Monte Carlo

Monte Carlo algorithms can be conceived based on the idea that the conditions on the right-hand side of the Wigner–Boltzmann equation represent terms for density gain and loss in the phase space. Let us begin with the semi-classical Boltzmann equation:

$$\left(\frac{\partial}{\partial t} + v(k).\nabla_r + \frac{1}{h}F(r).\nabla_k\right) f(r,k,t) = \\ \int dk' f(r,k',t) S(k',k) - \lambda\,(\mathrm{k}) f(r,k,t) \quad [3.52]$$

Here, S (k’,k) represents the rate of transition of the initial state k' to the final state k, induced by the physical collision process, and λ is the total diffusion rate:

$$\lambda\,(\mathrm{k}) = \int \mathrm{dk'S(k,k')}) \quad [3.53]$$

The negative term on the right-hand side of [3.54] is located in space k. In the Monte Carlo algorithm, the term $-\lambda_f$ gives rise to exponential distribution of free flight time of the carrier. The Wigner–Boltzmann equation has the same structure as that of [3.38]. We use [3.40] and increase it using Boltzmann collision operators:

$$\left(\frac{\partial}{\partial t} + v(k).\nabla_r + \frac{1}{h}F_{cl}(r).\nabla_k\right) f_w(r,k,t) = \int dk' \Gamma(k,k')\mu(k')f_w(r,k',t) - \mu(k)f_w(r,k,t) \quad [3.54]$$

Integral Γ takes the form of:

$$\Gamma(r,k,k') = \frac{1}{\mu(r,k')}[S(k',k) + V_m(r,k-k') + \alpha(k,r)\delta(k-k')] \quad [3.55]$$

$$\mu(r,k') = \lambda\,(\mathrm{r},\mathrm{k}') + \alpha(r,k') \quad [3.56]$$

with Γ being the normalization factor, it follows:

μ: normalization factor

$$\int dk'\Gamma(k,k',r) = 1 \quad [3.57]$$

$$S_{self}(k',k) = \alpha(r,k)\delta(k-k') \quad [3.58]$$

This imaginary mechanism of self-collision is introduced. Mathematically, contributions related to terms of gain and loss simply cancel out and hence have no effect. Physically, due to the Dirac function δ, the electron state of this mechanism does not change and hence the mechanism does not modify the trajectory of free flight. The choice of α provided a degree of freedom in constructing the Monte Carlo algorithm, as illustrated in what follows. Finally, we can conclude that the Monte Carlo methods seem to be well adapted to the study of devices (DGMOS), when compared with numerical methods that are often unstable in a dynamic state.

3.3.1. *Wigner–Boltzmann equation*

Equation [3.31] can be transformed into a path integral equation. The nucleus of the adjoint integral equation is:

$$P(k_f, t_f | k_i, t_i) = \Gamma[k_f, K(t_f)]\mu[K(t_f)]exp\left\{-\int_{t_i}^{t_f} \mu[K(\tau)]d\tau\right\} \quad [3.59]$$

The kernel represents a transition made of a free flight from time t_i and initial state k_i, followed by the final state collision k_f with the final time t_f. For the sake of brevity, dependency in space Γ and μ is omitted. In a Monte Carlo simulation, the duration of the next collisional event, T_f, is calculated from the exponential distribution in [3.46]:

$$p_t(t_f, t_i, k_i) = \mu[K(t_f)]exp\left\{-\int_{t_i}^{t_f} \mu[K(\tau)]d\tau\right\} \quad [3.60]$$

We note k', which is the state at the end of free flight, where k' = K (t_f). Transition of the end point of the trajectory k to the final state k'_f is carried out using nucleus Γ. Unlike the classical case, where P represents a transition probability, such an interpretation is not possible in the case of the Wigner equation, where P is not positive semi-definite. The problem arises from the Wigner potential, which takes positive and negative values. Owing to its anti-symmetrical properties with respect to q, the Wigner potential can be reformulated in terms of the positive function V^+:

$$V_w^+(r,q) = max(0, V_w(r,q)) \quad [3.61]$$

$$V_w(r,q) = V_w^+(r,q) - V_w^+(r,-q) \quad [3.62]$$

Then, the nucleus Γ is rewritten as a conditional probability distribution sum as follows:

$$\Gamma(k,k') = \frac{\lambda}{\mu} s(k',k) + \frac{\alpha}{\mu}\delta(k'-k) + \frac{\gamma}{\mu}[w(k,k') - w^*(k,k')] \quad [3.63]$$

$$s(k',k) = \frac{S(k',k)}{\lambda(k')}, w(k,k') = \frac{V_w^+(k-k')}{\gamma}, w^*(k,k') = w(k',k) \quad [3.64]$$

The normalizing factor for the Wigner potential is

$$\gamma(r) = \int dqV^+(r,q) \quad [3.65]$$

3.3.2. *Schrödinger–Poisson model*

A model based on the Poisson and Schrödinger equations is chosen to determine the electrical characteristics of the device. Coupling of the Poisson and Schrödinger equations is necessary when the oxide thickness and the height of the canal are close to the wavelength of electrons. This approach is justified by several results presented in the literature [BOH 51, TSU 73, CHA 74, THO 96, SVI 02, BOU 13, FIS 99].

3.3.2.1. *Poisson equation*

Integration of the Poisson equation makes it possible to calculate the variation in potential in the semiconductor. The space charge is made of charges created by carriers and free impurities that are assumed to be completely ionized (air). Recall that in two dimensions, the Poisson equation can be written as:

$$\frac{d^2V(x,y)}{dx^2} + \frac{d^2V(x,y)}{dy^2} = -\frac{q}{\varepsilon}[p - n + N_D - N_A] \quad [3.66]$$

where:

– V is the electric potential;

– p is the hole concentration;

– n is the electron concentration;

– N_D and N_A are donor and acceptor concentrations;

– q is the elementary charge;

– ε is the dielectric constant.

It is important to think about the possibility of using the Coulomb law instead of the Poisson equation, especially for very small devices; it will allow for more precision in the model and in calculation.

3.3.2.2. *Schrödinger equation*

Solving the Schrödinger equation makes it possible to obtain more energy levels and wave functions along the quantum confinement (along the y axis) if the potential variations are known beforehand. The Schrödinger equation to be solved is given by:

$$\frac{\hbar^2}{2m^*}\frac{\partial^2}{\partial y^2}\psi(y) + qV(y)\psi(y) = E\psi(y) \quad [3.67]$$

where ψ (y) is the wavefunction, E is the quantum energy, h is the Planck constant and V (y) is the electrostatic potential.

We can see that equations [3.54] and [3.55] are independent. We can illustrate a self-coherent system as follows:

$$\begin{cases} \rho(y) = S[V(y)] \\ V(y) = P[\rho(y)] \end{cases} \quad [3.68]$$

where functions S [V(y)] and P [ρ(y)] represent the Poisson and Schrödinger equations, respectively.

3.3.2.3. *Discretization method*

There are several discretization methods to solve coupled equations (Poisson–Schrödinger). Finite difference and finite element methods are widely used. We have adapted the finite difference method using a constant rectangular mesh, with the geometry of the device examined also being made of rectangles. The general equation to be discretized is of the following form:

$$\frac{\partial}{\partial x}\left[P(x,y).\frac{\partial U}{\partial x}\right]+\frac{\partial}{\partial y}\left[P(x,y).\frac{\partial U}{\partial y}\right]=f(U,x,y) \quad [3.69]$$

where:

– U is the unknown function;

– P and f are determined functions *a priori.*

Considering the integration of equation [3.57] around a point k (Figure 3.10: mesh cell) and a Green transform, we can discretize as follows [SHI 01]:

$$G_k.U_{k-1}+B_k.U_{k-n}+D_k.U_{k+1}+H_k.U_{k+1}-C_k.U_k=f(U_k,x_i,y_j) \quad [3.70]$$

where:

– i is the index along the x-axis;

– j is the index along the y-axis;

– k is the global index defined by:

$$\mathbf{k}=(\mathbf{j}-\mathbf{i})*\mathbf{n}+\mathbf{1}. \quad [3.71]$$

$$D_k=P\left(\frac{x_i+x_{i+1}}{2},y_j\right).\frac{2}{(x_{i+1}-x_i).(x_{i+1}-x_{i-1})}$$

$$G_k=P\left(\frac{x_i+x_{i-1}}{2},y_j\right).\frac{2}{(x_i-x_{i-1}).(x_{i+1}-x_{i-1})}$$

$$B_k=P\left(x_i,\frac{y_j+y_{j-1}}{2}\right).\frac{2}{(y_j-y_{j-1}).(y_{j+1}-y_{j-1})}$$

$$H_k=P\left(x_i,\frac{y_j+y_{j+1}}{2}\right).\frac{2}{(y_{j+1}-y_j).(y_{j+1}-y_{j-1})}$$

$$C_k=G_k+D_k+B_k+H_k$$

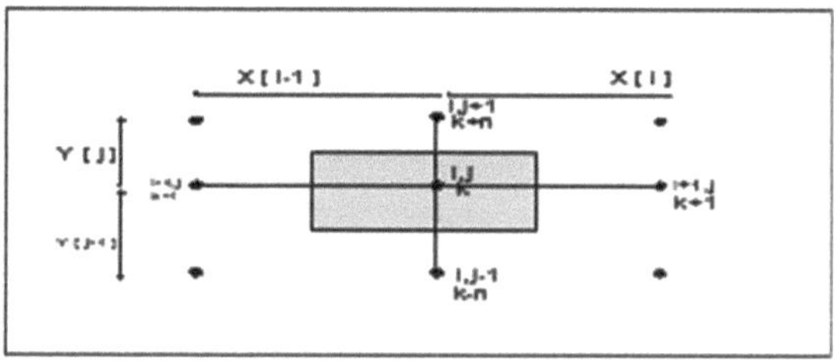

Figure 3.45. *Mesh cell*

3.3.2.4. *FDM discretization*

The calculation domain is ($N_X \times N_Y$) nodes, where $N_X \times N_Y$ are the number of nodes in x and y directions, respectively. The numerical solution of the Poisson equation is composed of potential values. The same number of equations is required to obtain the $N_X \times N_Y$ unknown values.

By applying FDM (Finite Difference Method) to equation [3.54], we can write [SIL]:

$$\frac{\Delta y}{\Delta x}V_{i-1,j} + \frac{\Delta x}{\Delta y}V_{i,j-1} - 2\left(\frac{\Delta x}{\Delta y} + \frac{\Delta x}{\Delta y}\right)V_{i,j} + \frac{\Delta x}{\Delta y}V_{i,j+1} + \frac{\Delta y}{\Delta x}V_{i+1,j} = -\Delta x.\Delta y - \frac{q}{\varepsilon}[n + N_D - N_A] \quad [3.72]$$

where Δx and Δy are mesh dimensions in the x and y directions, respectively. The discretized Poisson equation can be written in the matrix form: S = M.X, where M is a matrix and G, B, D, H and C are coefficients:

$$[M] = \begin{bmatrix} C\ D \ldots H \ldots \\ G\ C\ D \ldots H .. \\ .G\ C\ D \ldots H . \\ ..G\ C\ D \ldots H \\ \ldots G\ C\ D \ldots \\ B \ldots G\ C\ D .. \\ .B \ldots G\ C\ D . \\ ..B \ldots G\ C\ D \\ ..B \ldots G\ D \end{bmatrix} \quad [3.73]$$

with coefficients $G_k - D_k = \frac{\Delta y}{\Delta x}$, $B_k = H_k = \frac{\Delta x}{\Delta y}$

When nodes are at the silicon/oxide interface (without charge), the dielectric constant varies and the displacement vector continuity is analyzed, equation [3.59] becomes [ORI 99]:

$$\frac{\Delta y}{\Delta x}V_{i-1,j}+\frac{\Delta x}{2\Delta y}\left(1+\frac{\varepsilon_{bot}}{\varepsilon_{top}}\right)V_{i,j-1}-\left(\frac{\Delta x}{\Delta y}+\frac{\Delta x}{\Delta y}\right)\left(1+\frac{\varepsilon_{bot}}{\varepsilon_{top}}\right)V_{i,j}+\frac{\Delta x}{2\Delta y}\left(1+\frac{\varepsilon_{bot}}{\varepsilon_{top}}\right)V_{i,j+1}$$
$$+\frac{\Delta y}{\Delta x}\frac{\varepsilon_{bot}}{\varepsilon_{top}}V_{i+1,j}=-\Delta x.\Delta y-\frac{q}{\varepsilon_{top}}[n+N_D-N_A] \quad [3.74]$$

where ε_{bot} and ε_{top} are dielectric constants of materials above and below the interface. Here, both oxides are identical.

3.3.2.5. *Boundary conditions*

To solve these equations, boundary conditions must be used ("Dirichlet" and "Newman").

3.3.2.6. *Contact conditions*

Given N_D, N_A and n, equations [3.60]–[3.69] represent a linear equation set that can be directly solved using an iterative method. However, when solving a coupled equation set, there is a better algorithm for solving the Poisson equation [DAT 00, SVI 02, THO 96]. This algorithm involves substituting variable n. The potential energy of the quasi-Fermi level is calculated according to the old potential:

$$(F_n)_{i,j} = -q(V_{old})_{i,j} + k_B T.\Im_{1/2}^{-1}\left(\frac{n_{i,j}}{N_C}\right) \quad [3.75]$$

where:

– $\Im_{1/2}^{-1}$ is the Fermi integral of order ½;

– N_C is the density of state in the conduction band.

The density of electrons is therefore:

$$n_{i,j} = N_C\Im_{1/2}\left[\frac{(F_n)_{i,j}+qV_{i,j}}{k_B T}\right] \quad [3.76]$$

The Poisson equation is then solved using the Newton–Raphson method:

$$F(\vec{V_i}) = 0 \quad [3.77]$$

3.3.2.7. *Solving the Schrödinger equation*

In the concerned DGMOS, the confinement can be considered to be efficient only in one direction; the Schrödinger equation is therefore solved in one dimension.

To solve the Schrödinger equation, the structure is divided into two parts, called Schrödinger parts, along the transport. In each part (or column), we know the potential profile by solving the Poisson equation in the same way as solving the Schrödinger equation for confinement (y-axis) made it possible to obtain the energy levels and wave functions [CLE 01]:

$$-\frac{\hbar^2}{2m^*}\frac{\partial^2}{\partial y^2}\psi_i(y) + qV(y)\psi_i(y) = E_i\psi_i(y) \qquad [3.78]$$

$$-\frac{\hbar^2}{2m^*(y)}\frac{\psi_i(y_{i+1})-2\psi_i(y_i)+\psi_i(y_{i-1})}{\Delta y^2} + qV(y_i)\psi_i(y_i) = E_i\psi_i(y_i) \qquad [3.79]$$

$$j{=}o;\ -\frac{\hbar^2}{2m^*(y)}\frac{\psi_i(y_1)-2\psi_i(y_o)+\psi_i(y_{-1})}{\Delta y^2} + qV(y_o)\psi_i(y_{io}) = E_i\psi_i(y_o) \qquad [3.80]$$

$$j{=}1;-\frac{\hbar^2}{2m^*(y)}\frac{\psi_i(y_2)-2\psi_i(y_1)+\psi_i(y_o)}{\Delta y^2} + qV(y_1)\psi_i(y_1) = E_i\psi_i(y_1) \qquad [3.81]$$

$$j{=}N;-\frac{\hbar^2}{2m^*(y)}\frac{\psi_i(y_{N+1})-2\psi_i(y_N)+\psi_i(y_{N-1})}{\Delta y^2} + qV(y_N)\psi_i(y_N) = E_i\psi_i(y_N) \qquad [3.82]$$

$$\text{where } (y) = E_i\psi_i(y) \qquad [3.83]$$

and H [BAR 99]:

$$H = -\frac{h^2}{2m^*(y)}\begin{bmatrix} -2 & 1 & 0\ \dots & \dots & 0 \\ 1 & -2 & 1\ 0 & \dots & 0 \\ 0 & 1 & -2\ 1 & \dots & 0 \\ \dots & \dots & \dots\dots & \dots & \dots \\ 0 & \dots & 0\ 1 & -2 & 1 \\ 0 & \dots & \dots 0 & 1 & -2 \end{bmatrix} + q\begin{bmatrix} V(y_1) & 0 & \dots\dots & \dots & 0 \\ 0 & V(y_2) & 0\dots & \dots & 0 \\ \dots & \dots & \dots\dots & \dots & \dots \\ 0\ \dots & 0\,V(y_i) & 0 & & \dots \\ \dots\ \dots & \dots\ \dots & \dots & & \dots \\ 0\ \dots & \dots\ \dots & \dots & & V(y_N) \end{bmatrix} \qquad [3.84]$$

3.3.3. *Numerical results*

First, we compare the electrical characteristics obtained by classic and quantum methods. The current level in quantum structures is weaker than that observed in classic structures [TSU 73]. Figure 3.46 illustrates the drain current for Vds = 0.6 V and shows the difference between the two models.

This is due primarily to the differences existing in the density of carriers in the active zone [CHA 74]. If the transistor dimensions are below a few tens of nanometers, known physical laws come into play.

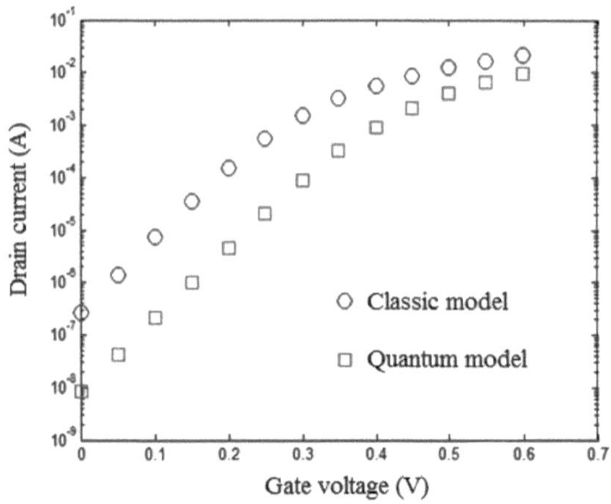

Figure 3.46. *(I_D–V_{GS}) characteristics: comparison of the classic and quantum models*

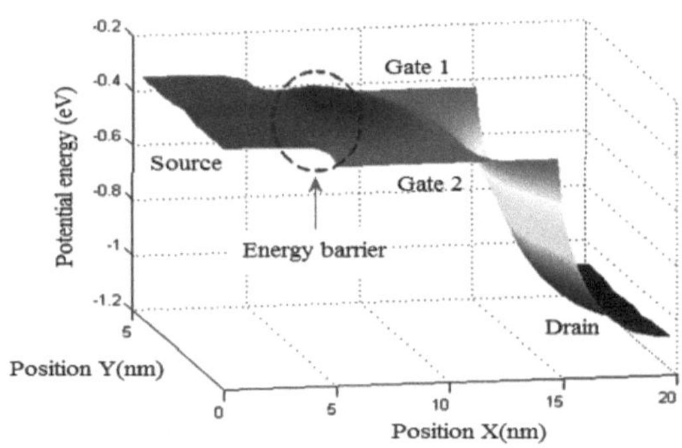

Figure 3.47. *Potential energy of DGMOS. For a color version of this figure, see www.iste.co.uk/gontrand/nanoelectronics.zip*

3.3.4. *Short-channel effects*

The DGMOSFET (DGMOS) emerged as the most promising device for conceiving nanometric circuits. This is primarily due to its excellent control of short-channel effects. These effects are controlled in several ways, such as leakage current (I_{OFF}), tunneling effect and drain-induced barrier lowering (DIBL).

3.3.4.1. *DIBL*

The analysis of I_D (V_{DS}) characteristics help identify the DIBL for different channel lengths for low and high drain polarizations. Figure 3.48 illustrates the effects of channel length variation on the DIBL for a weak (Vds = 0.1 V) and high (Vds = 1.5 V) polarization. We can observe that when the gate work function increases from 4.25 to 4.5 eV, the DIBL is weaker. Moreover, the DIBL is only slightly affected by the gate work function for channel lengths higher than 12 nm.

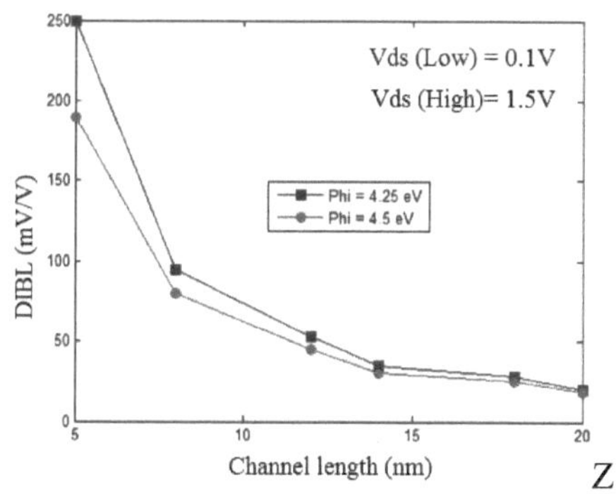

Figure 3.48. *DIBL versus channel length for two gate work functions*

3.3.4.2. *Tunneling effect on the DGMOS structure*

Another parameter characterizing performance of the short channel is the tunneling current. In MOS structures, the tunneling effect is characterized by carriers moving through the oxide. There are three types of tunneling effect [SVI 02]:

– direct tunneling current is dominant in conduction in ultra-thin oxide MOS structures (Tox < 3 nm) working at a voltage lower than 2 V;

– indirect tunneling current exists in such structures, but only in the case of very thin oxide, indirect tunneling current is negligible when compared with direct tunneling current;

– Fowler–Nordheim tunneling current appears in MOS structures in which the oxide thickness is greater than 3 nm or less when the polarization voltages are high.

In our case, the oxide thickness is equal to 1.5 nm and voltages applied are below 2 V. The continuous gate current is calculated using:

$$J_G = \sum_j \sum_i f_{i,j}\, Q_{i,j}\, T_{TD}\left(E_{i,j}\right) \qquad [3.85]$$

where:

– $f_{i,j}$ is the estimated frequency impact of electrons striking the oxide barrier;

– $T_{TD}(E_{i,j})$ is the tunnel transparency;

– $Q_{i,j}$ is the total charge.

Tunnel transparency

Tunnel transparency is defined as the relationship of the transmitted current to the incident current, relating to incident and transmitted waves [GHA 88]. To calculate the transparency of all types of tunneling barriers in order to obtain analytical expressions, we take the WKB (Wentzel–Kramers–Brillouin) approximation. The tunnel transparency can then be written under the form of equation [BOU 13]:

$$T_{WKB}\left(E_{i,j}\right) = \exp\left[-\frac{8\pi}{3}\frac{T_{OX}\sqrt{2m_{OX}}}{qV_{OX}h}\left(\left(q\Phi_{OX}-E_{i,j}\right)^{3/2} - \left(q\Phi_{OX}-qV_{OX}-E_{i,j}\right)^{3/2}\right)\right] \quad [3.86]$$

where:

– T_{ox} is the oxide thickness;

– m_{ox} is the electron mass in oxide;

– Φ_{ox}: potential barrier height for electrons in the oxide;

– $E_{i,j}$ is the potential energy at the oxide/semiconductor interface;

– V_{OX} is the potential in oxide.

Frequency of impact

We assume that the frequency f of the impact of electrons is limited by the movement within the semiconductor experiencing the tunneling effect both from the substrate to the gate and from the gate to the substrate. The impact frequency can be expressed as [BOU 13]:

$$f_{i,j} = \frac{q\varepsilon_{OX}V_{OX}}{4T_{OX}\varepsilon_S}\left(\frac{m_j\ E_{i,j}}{3}\right)^{-1/2} \quad [3.87]$$

where:

– ε_{ox} is the permittivity of oxide;

– ε_s is the permittivity of the semiconductor.

Total charge

The total charge is given by:

$$Q_{i,j} = q \frac{g_j \; m_j}{\pi \; \bar{h}^2} K_B T \; \ln \left[1 + \exp \left(\frac{E_{Fn} - E_{i,j}}{K_B T} \right) \right] \qquad [3.88]$$

In the concerned structure (Figure 3.49), there are two levels of tunnel current: I_{G1} and I_{G2}.

The total tunnel current is their sum:

$$I_G = I_{G1} + I_{G2} \qquad [3.89]$$

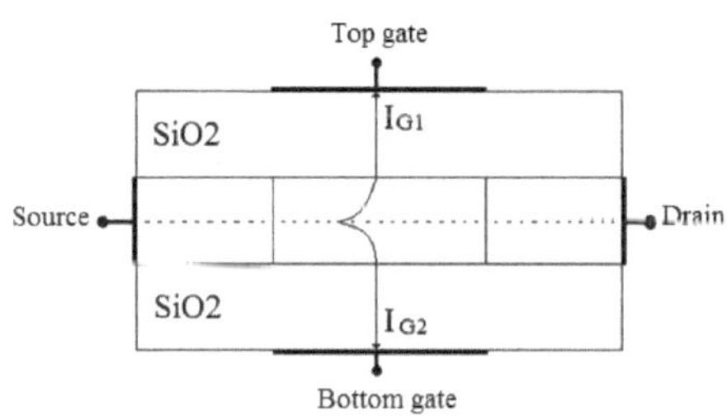

Figure 3.49. *Tunneling currents in a DGMOSFET*

In Figure 3.52, we observe the effect of gate and drain voltage of the DGMOS. We observe that the drain current increases when the gate current decreases; while the drain voltage increases, the gate current is proportional to the transverse electric field.

3.4. Electromagnetism; MOS: Maxwell equations

3.4.1. *Introduction*

A detailed electromagnetic analysis is required to understand high-frequency circuits. It concerns microwave devices and nanoelectronics. The method used is a global electromagnetic analysis in the spatial and temporal domains.

In this section, we will first highlight the electromagnetic model chosen for our study. The model is based on Maxwell equations.

We will see the FDTD (Finite Difference Time Domain) method, which is a numerical approach that makes it possible to solve differential equations in the spatial and temporal domains.

Yee was the first to apply this method to Maxwell equations in free space in 1966 [YEE 66]. This method involves approaching spatial and temporal partial derivatives that come up in Maxwell equations using centered finite differences.

Thus, we will underline key points of the FDTD such as discretization of Maxwell equations using centered finite differences, stability criterion due to the choice of time step and numerical dispersion due to the choice of spatial mesh. We will also elaborate on dealing with certain boundary conditions that will be used in this paper, such as the PEC (Perfect Electric Conductor) type and PMC (Perfect Magnetic Conductor) type. Moreover, to delimit the computational space and simulate an infinite open space around the structure, UPML (Uniaxial Perfectly Matched Layer) absorbing conditions are used [TAF 05].

Furthermore, the FDTD method, in its basic formulation, does not take into account the presence of localized elements (linear and nonlinear) in the computational space. This issue can be solved by modifying the Maxwell–Ampère equation such that the method encompasses LE-FDTD (Lumped Element-FDTD) localized elements [WEN 06, GED 96].

We will thus make numerical simulations relating to interconnections (microstrip line) to show their effects on the working of microwave circuits. This experiment will be carried out on a large frequency band, reaching up to 100 GHz.

Finally, we will simulate a nonlinear localized element such as MOS RF in the temporal domain using a calculation code developed using MATLAB.

3.4.2. *Maxwell equations*

Let us consider an area in space lacking electric and magnetic sources. Using the international system of units, generalized Maxwell equations are written in their local integral and differential forms with the laws given below [LAB 11].

– Faraday's law:

$$\frac{\partial \vec{B}}{\partial t} = -\vec{\nabla} \times \vec{E} - \vec{J}_m$$

$$\frac{\partial}{\partial t}\iint_S \vec{B}.d\vec{S} = -\oint_C \vec{E}.d\vec{l} - \iint_S \vec{J}_m.d\vec{S} \qquad [3.90]$$

– Ampere's law:

$$\frac{\partial \vec{D}}{\partial t} = \vec{\nabla} \times \vec{H} - \vec{J}_e$$

$$\frac{\partial}{\partial t}\iint_S \vec{D}.d\vec{S} = -\oint_C \vec{H}.d\vec{l} - \iint_S \vec{J}_e.d\vec{S} \qquad [3.91]$$

– Gauss' law for the electrical field (in the absence of an electrical load):

$$\vec{\nabla}.\vec{D} = 0$$

$$\oiint_S \vec{D}.d\vec{S} - 0 \qquad [3.92]$$

Gauss' law for magnetic field (in the absence of a magnetic load):

$$\vec{\nabla}.\vec{B} = 0$$

$$\oiint_S \vec{B}.d\vec{S} = 0 \qquad [3.93]$$

In a homogeneous, isotropic and non-dispersive environment, we have:

$$\vec{B} = \mu.\vec{H} \qquad [3.94]$$

$$\vec{D} = \varepsilon.\vec{E} \qquad [3.95]$$

where:

– E is the electric field (V/m);

– D is the electrical excitation (C/m^2);

– H is the magnetic excitation (A/m),

– B is the magnetic field (Tesla);

– J_e is the electrical conduction current density (A/m^2);

– J_m is the magnetic conduction current density (A/m^2);

– μ is the magnetic permeability (H/m);

– ε is the electric permittivity (F/m).

If σ is the electrical conductivity and σ_m the magnetic conductivity, we obtain expressions for J_m and J_e according to E and H as:

$$\vec{J}_m = \sigma_m.\vec{H} \qquad [3.96]$$

$$\vec{J}_e = \sigma.\vec{E} \qquad [3.97]$$

3.4.3. *Overview of the FDTD method*

The FDTD method makes it possible to solve Maxwell equations with finite differences in the temporal domain. This method requires double discretization: spatial and temporal.

This leads to the construction of the structure's mesh that needs to be studied as well as to discretization of the temporal axis. By requiring excitation in one point on the gate, devices of the electromagnetic field will be calculated iteratively at all points on the gate and at all times. Temporal evolution of the electromagnetic field can be determined at any point on the gate. A broadband excitation source can help characterize linear structures on a broadband frequency in one simulation.

3.4.3.1. *Principle of the Yee method*

We take a modeling space (*O, x, y, z*) into account. Δx, Δy and Δz represent discretization steps in three directions and *i, j* and *k* the coordinates of a particular point in space.

When $J_m=J_e=0$, the projection of two equations [3.1] and [3.2] on three coordinate axes *x, y* and *z* is:

$\frac{\partial H_x}{\partial t} = -\frac{1}{\mu}\left(\frac{\partial E_z}{\partial y} - \frac{\partial E_y}{\partial z}\right)$	$\frac{\partial H_y}{\partial t} = -\frac{1}{\mu}\left(\frac{\partial E_x}{\partial z} - \frac{\partial E_z}{\partial x}\right)$	$\frac{\partial H_z}{\partial t} = -\frac{1}{\mu}\left(\frac{\partial E_y}{\partial x} - \frac{\partial E_x}{\partial y}\right)$
$\frac{\partial E_x}{\partial t} = \frac{1}{\varepsilon}\left(\frac{\partial H_z}{\partial y} - \frac{\partial E_y}{\partial z}\right)$	$\frac{\partial E_y}{\partial t} = \frac{1}{\varepsilon}\left(\frac{\partial H_x}{\partial z} - \frac{\partial E_z}{\partial x}\right)$	$\frac{\partial E_z}{\partial t} = \frac{1}{\varepsilon}\left(\frac{\partial H_y}{\partial x} - \frac{\partial E_x}{\partial y}\right)$

In order to use these equations in an algorithm for numerical calculations, the approximation of centered derivative and the Yee leapfrog diagram [YEE 66, MIR 10] (smoothing effect) illustrated in Figure 3.50 will be used, which lead to the following system of equations:

$$E_x^{n+1}(i,j,k) = E_x^n(i,j,k) + \frac{dt}{\varepsilon.\Delta y}\left[H_z^{n+\frac{1}{2}}(i,j,k) - H_z^{n+\frac{1}{2}}(i,j-1,k)\right] - \frac{dt}{\varepsilon.\Delta z}\left[H_y^{n+\frac{1}{2}}(i,j,k) - H_y^{n+\frac{1}{2}}(i,j,k-1)\right] \quad [3.98]$$

$$E_y^{n+1}(i,j,k) = E_y^n(i,j,k) + \frac{dt}{\varepsilon.\Delta z}\left[H_x^{n+\frac{1}{2}}(i,j,k) - H_x^{n+\frac{1}{2}}(i,j,k-1)\right] - \frac{dt}{\varepsilon.\Delta x}\left[H_z^{n+\frac{1}{2}}(i,j,k) - H_z^{n+\frac{1}{2}}(i-1,j,k)\right] \quad [3.99]$$

$$E_z^{n+1}(i,j,k) = E_z^n(i,j,k) + \frac{dt}{\varepsilon.\Delta x}\left[H_y^{n+\frac{1}{2}}(i,j,k) - H_y^{n+\frac{1}{2}}(i-1,j,k)\right] - \frac{dt}{\varepsilon.\Delta y}\left[H_x^{n+\frac{1}{2}}(i,j,k) - H_x^{n+\frac{1}{2}}(i,j-1,k)\right] \quad [3.100]$$

$$H_x^{n+\frac{1}{2}}(i,j,k) = H_x^{n-\frac{1}{2}}(i,j,k) + \frac{dt}{\mu.\Delta z}\left[E_y^n(i,j,k+1) - E_y^n(i,j,k)\right] - \frac{dt}{\mu.\Delta y}\left[E_z^n(i,j+1,k) - E_z^n(i,j,k)\right] \quad [3.101]$$

$$H_y^{n+\frac{1}{2}}(i,j,k) = H_y^{n-\frac{1}{2}}(i,j,k) + \frac{dt}{\mu.\Delta x}\left[E_z^n(i+1,j,k) - E_z^n(i,j,k)\right] - \frac{dt}{\mu.\Delta z}\left[E_x^n(i,j,k+1) - E_x^n(i,j,k)\right] \quad [3.102]$$

$$H_z^{n+\frac{1}{2}}(i,j,k) = H_z^{n-\frac{1}{2}}(i,j,k) + \frac{dt}{\mu.\Delta x}y\left[E_z^n(i,j+1,k) - E_x^n(i,j,k)\right] - \frac{dt}{\mu.\Delta x}\left[E_y^n(i+1,j,k) - E_y^n(i,j,k)\right] \quad [3.103]$$

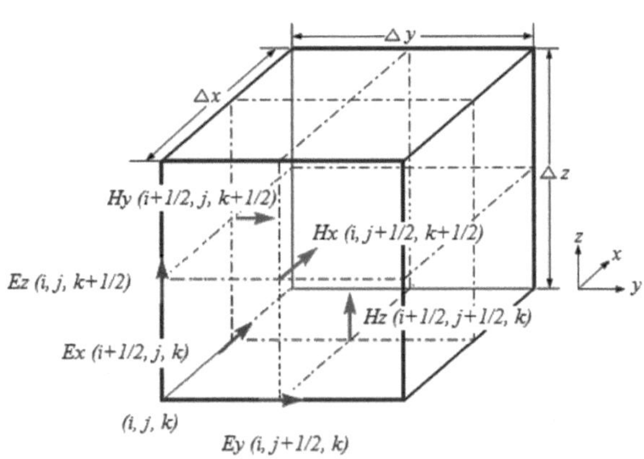

Figure 3.50. *The Yee cell. For a color version of this figure, see www.iste.co.uk/gontrand/nanoelectronics.zip*

Electromagnetic source

To study the characteristics of the structure taken into account, the structure must be electromagnetically excited. The choice of electromagnetic source will depend on the form of the structure and the targeted frequency band.

Stability criterion

The numerical stability criterion fulfills two requirements: spatial and numerical. In fact, the values of the field at all points of the computational volume depend on the values obtained at the preceding times. This schema operates under a stability criterion linking the time step Δt to spatial steps Δx, Δy and Δz. Owing to the stability criterion, numerical sampling should be enough to follow the temporal evolution of the electromagnetic field. This has been shown using the "Friedrichs–Lewy" condition:

$$\Delta t \leq \frac{1}{c\sqrt{\frac{1}{\Delta x^2}+\frac{1}{\Delta y^2}+\frac{1}{\Delta z^2}}} \quad [3.104]$$

where the speed of light in free space is ($\sim 3.10^{+8}$ m/s).

This convergence criterion is defined in the case of a uniform mesh. However, the same criterion can be retained in another type of orthogonal mesh, for instance a non-uniform one.

Numerical dispersion

When an electromagnetic signal propagates in a meshed computational domain using finite differences, it undergoes transformations (distortion, attenuation) due, among other things, to mesh dispersive effects. These effects are not only due to discretization that provides an approximate representation of signals, but also, and more importantly, due to details regarding the formulations used for partial derivatives. In other words, this dispersion will depend, first, on the cell size: the Δx, Δy and Δz cell size in comparison to the smallest wavelength present in the spectrum of analysis, and on the order of the error committed during evaluation of the partial derivatives in Maxwell equations [TAF 05, FIS 99].

In short, by choosing a spatial increment value less than $\lambda/10$ (λ: smaller wavelength), it is fair to say that the dispersion is negligible.

3.4.4. *Perfect electric conductor (PEC)*

In the case of a perfect conductor, tangential devices of the electric field must be forced to zero. In the case of real conductors, whose conductivity is known, we will assign it to elementary cells in the FDTD space.

According to Figure 3.51, the two perpendicular devices of the electric field are defined as:

$$E_x(i,j,k+1) = E_x(i,j+1,k+1) = 0$$

$$E_y(i,j,k+1) = E_y(i+1,j,k+1) = 0 \qquad [3.105]$$

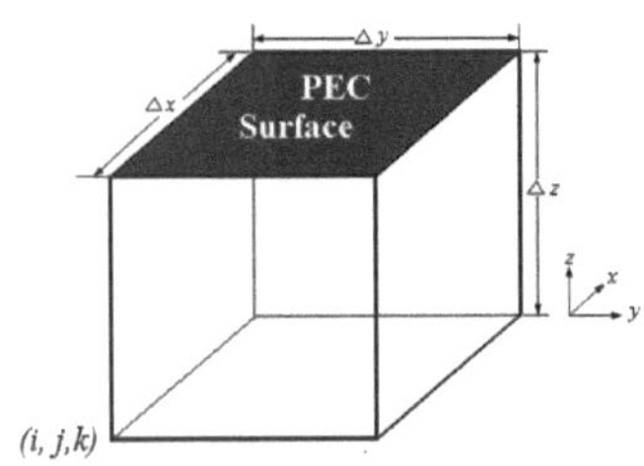

Figure 3.51. *PEC condition in the upper face of the cube (i, j, k). For a color version of this figure, see www.iste.co.uk/gontrand/nanoelectronics.zip*

Generally, this type of condition is applied in two cases:

– when the electrical conductivity is high and the skin effect is negligible;

– to define an antisymmetric plane of the electrical field.

3.4.5. *Perfect magnetic conductor (PMC)*

A boundary condition like the perfect magnetic conductor (PMC) requires that the tangential device of the electric field H on the concerned face be nullified.

According to Figure 3.52, the PMC boundary condition is defined as [LI 08]:

$$\vec{n} \times \vec{H} = \vec{0} \qquad [3.106]$$

with $\vec{n}$ being unit normal vector of the concerned face.

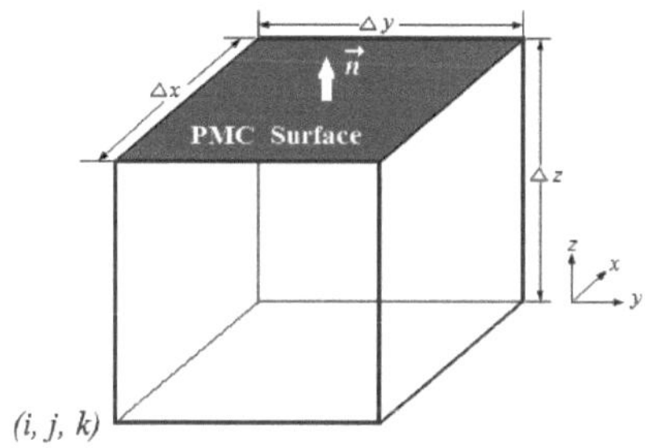

Figure 3.52. *PMC condition in the upper face of the cube (i, j, k). For a color version of this figure, see www.iste.co.uk/gontrand/nanoelectronics.zip*

Generally, this type of condition is applied in two cases:

– on a surface of a high permeability material;

– to define an antisymmetric plane of the magnetic field.

3.4.6. *Absorbing condition*

A huge challenge in simulating FDTD-3D is to deal with radiation problems using absorbing boundary conditions (ABC) [LI 08].

In such applications, boundaries of the domain must be defined as having infinite radiation; there is NO reflection. One possibility is the use of UPML (Uniaxial Perfectly Matched Layer) boundary conditions that allow for plane waves to be absorbed but require that the direction of propagation be known. We can also circumvent this problem by defining an anisotropic and artificial material that is transparent and infinitely absorbing (material having high loss) along the incident wave. Therefore, one UPML area is not a boundary condition, but rather a simulation artifice that allows for simulating infinite domains.

Maxwell equations on propagation in an anisotropic setting are defined for harmonic modes:

$$\vec{\nabla}\times\vec{H} = j.\omega.\varepsilon.\bar{\bar{s}}.\vec{E} \qquad [3.107]$$

$$\vec{\nabla}\times\vec{E} = -j.\omega.\mu.\bar{\bar{s}}.\vec{H} \qquad [3.108]$$

where $\vec{E}$ and $\vec{H}$ represent vectors of the electric and magnetic field with the phasor form and where s represents the diagonal tensor, given by:

$$\overline{\overline{s}} = \begin{bmatrix} \frac{s_y s_z}{s_x} & 0 & 0 \\ 0 & \frac{s_x s_z}{s_y} & 0 \\ 0 & 0 & \frac{s_x s_y}{s_z} \end{bmatrix} \qquad [3.109]$$

where s_x, s_y and s_z are complex permittivities in the x, y and z directions, which are given by:

$$\begin{cases} s_x = k_x + \dfrac{\sigma_x}{j\omega\varepsilon} \\ s_y = k_y + \dfrac{\sigma_y}{j\omega\varepsilon} \\ s_z = k_z + \dfrac{\sigma_z}{j\omega\varepsilon} \end{cases} \qquad [3.110]$$

In order to anticipate disadaptation problems between the isotropic setting and UMPL layers (Figure 3.4), and to thus reduce reflections, the conductivity of the UPML layer gradually varies all along the three axes x, y and z as in the following equations:

$$\begin{cases} \sigma_{x,y,z}(u) = \left(g^{\frac{1}{\Delta}} \right)^u .\sigma_0 \\ k_{x,y,z}(u) = \left(g^{\frac{1}{\Delta}} \right)^u \end{cases} \qquad [3.111]$$

where u is the normal distance of the UPML area, Δ is the spatial discretization step, and g and σ_0 are constants (conductivity of the material).

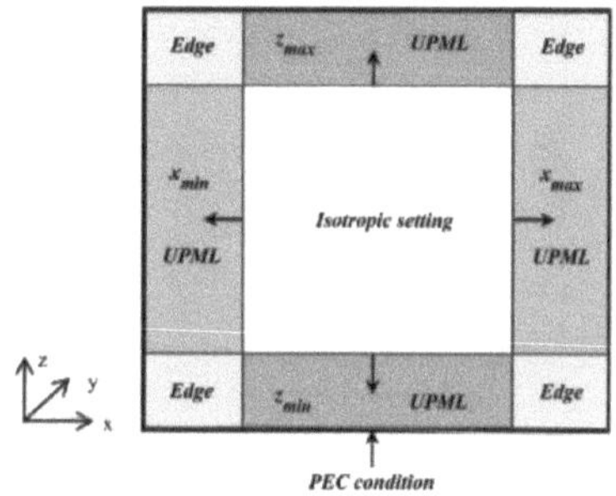

Figure 3.53. *UPML absorbing condition in the (x, z) plane. For a color version of this figure, see www.iste.co.uk/gontrand/nanoelectronics.zip*

The vector form of equation [3.28] is developed using the following expression:

$$\begin{bmatrix} \dfrac{\partial \bar{H}_z}{\partial y} - \dfrac{\partial \bar{H}_y}{\partial z} \\ \dfrac{\partial \bar{H}_x}{\partial z} - \dfrac{\partial \bar{H}_z}{\partial x} \\ \dfrac{\partial \bar{H}_y}{\partial x} - \dfrac{\partial \bar{H}_x}{\partial y} \end{bmatrix} = j\omega\varepsilon \begin{bmatrix} \dfrac{s_y s_z}{s_x} & 0 & 0 \\ 0 & \dfrac{s_x s_z}{s_y} & 0 \\ 0 & 0 & \dfrac{s_x s_y}{s_z} \end{bmatrix} . \begin{bmatrix} \bar{E}_x \\ \bar{E}_y \\ \bar{E}_z \end{bmatrix} \qquad [3.112]$$

In the UPML area, the electric field and the electric flux density are linked using the following three formulae:

$$\begin{cases} \bar{D}_x = \varepsilon \dfrac{s_z}{s_x} \bar{E}_x \\ \bar{D}_y = \varepsilon \dfrac{s_x}{s_y} \bar{E}_y \\ \bar{D}_z = \varepsilon \dfrac{s_y}{s_z} \bar{E}_z \end{cases} \qquad [3.113]$$

Substituting equation [3.34] into [3.43], a new formula can be obtained.

$$\begin{bmatrix} \dfrac{\partial \bar{H}_z}{\partial y} - \dfrac{\partial \bar{H}_y}{\partial z} \\ \dfrac{\partial \bar{H}_x}{\partial z} - \dfrac{\partial \bar{H}_z}{\partial x} \\ \dfrac{\partial \bar{H}_y}{\partial x} - \dfrac{\partial \bar{H}_x}{\partial y} \end{bmatrix} = j\omega \begin{bmatrix} s_y & 0 & 0 \\ 0 & s_z & 0 \\ 0 & 0 & s_x \end{bmatrix} . \begin{bmatrix} \bar{D}_x \\ \bar{D}_y \\ \bar{D}_z \end{bmatrix} \qquad [3.114]$$

Passing from the frequency space towards the temporal space can be carried out by substituting the term $j\omega$ with the operator $\frac{\partial}{\partial t}$.

$$\begin{bmatrix} \dfrac{\partial H_z}{\partial y} - \dfrac{\partial H_y}{\partial z} \\ \dfrac{\partial H_x}{\partial z} - \dfrac{\partial H_z}{\partial x} \\ \dfrac{\partial H_y}{\partial x} - \dfrac{\partial H_x}{\partial y} \end{bmatrix} = \frac{\partial}{\partial t} \begin{bmatrix} k_y & 0 & 0 \\ 0 & k_z & 0 \\ 0 & 0 & k_x \end{bmatrix} . \begin{bmatrix} D_x \\ D_y \\ D_z \end{bmatrix} + \frac{1}{\varepsilon} \begin{bmatrix} \sigma_y & 0 & 0 \\ 0 & \sigma_z & 0 \\ 0 & 0 & \sigma_x \end{bmatrix} . \begin{bmatrix} D_x \\ D_y \\ D_z \end{bmatrix} \qquad [3.115]$$

Using the same method for equation [3.29], the variation of the magnetic flux with time in the UPML area is given by:

$$\begin{bmatrix} \dfrac{\partial E_z}{\partial y} - \dfrac{\partial E_y}{\partial z} \\ \dfrac{\partial E_x}{\partial z} - \dfrac{\partial E_z}{\partial x} \\ \dfrac{\partial E_y}{\partial x} - \dfrac{\partial E_x}{\partial y} \end{bmatrix} = -\frac{\partial}{\partial t} \begin{bmatrix} k_y & 0 & 0 \\ 0 & k_z & 0 \\ 0 & 0 & k_x \end{bmatrix} . \begin{bmatrix} B_x \\ B_y \\ B_z \end{bmatrix} + \frac{1}{\varepsilon} \begin{bmatrix} \sigma_y & 0 & 0 \\ 0 & \sigma_z & 0 \\ 0 & 0 & \sigma_x \end{bmatrix} . \begin{bmatrix} B_x \\ B_y \\ B_z \end{bmatrix} \qquad [3.116]$$

3.4.7. *Modeling localized linear equations*

The FDTD method transforms a spatio-temporal discretization of Maxwell equations. This method takes into account all electromagnetic phenomena (propagation, reflection, diffraction, coupling, etc.) in an electronic circuit.

However, its basic form cannot integrate localized elements in a computational volume. In fact, the FDTD is based on iterative calculation of devices of an electric field E and magnetic field H, while localized linear or nonlinear elements are modeled using voltage–current (*I*–*V*)-type electric equations.

This technique makes it possible to simply consider circuit models that are not only closer to reality but also complex. Several diode and transistor models can be implemented in the FDTD algorithm in order to model several complex microwave circuits [GON 06].

Moreover, inserting nonlinear elements can be improved by dynamically adjusting the time step. This stabilizes the algorithm, while making it converge quickly [MIR 10].

Coupling between the FDTD algorithm and the circuit model occurs through either equivalent sources or thin wire formalism. In general, the temporal method of solving Maxwell equations is coupled with a simulator that solves Kirchhoff laws in a temporal domain such as SPICE (Simulation Program for Integration Circuit Emphasis). These coupling techniques described in [QUA 07] and [TAF 05] made it possible, for instance, to model active antennas.

Inserting localized elements can be done simply in a formulation based on the Maxwell–Ampere equation (equations [2.4.1]). In an elementary cell (Figure 3.44), this equation changes the relation between the total current flowing through $\nabla \times H$ cell and conduction $Jc = \sigma E$ and displacement $\varepsilon.\partial E/\partial t$ currents.

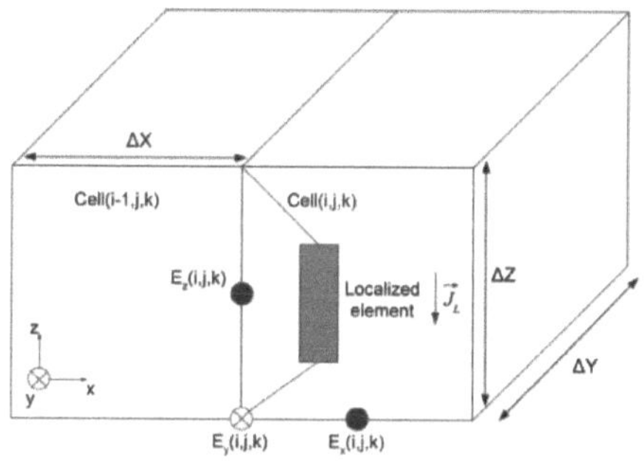

Figure 3.54. *Insertion of a localized element in an FDTD gate*

In the Maxwell–Ampere equation, a term for current density J_L, modeling local current circulating in the localized element, can be added to conduction and displacement current densities [CIA 96]:

$$\vec{\nabla}\times\vec{H}=\varepsilon\frac{\partial\vec{E}}{\partial t}+\vec{J}_C+\vec{J}_L \qquad [3.117]$$

The current density J_L is linked to the current circulating in the localized elements, thanks to equation [3.118], with the current flowing through the surface of the elementary cell (*ΔXΔY*).

$$J_L=\frac{I_L}{\Delta X.\Delta Y} \qquad [3.118]$$

The current I_L is more or less a complex function of the voltage V linked to the field E:

$$V=-E_Z(i,j,k).\Delta Z \qquad [3.119]$$

Inserting passive and active localized elements in the FDTD was carried out in two dimensions. This method was improved to integrate 3D localized elements through several cells.

3.4.7.1. *In the case of resistance*

Based on Ohm's law, V=R.I, the density of the current circulating in the resistance can be expressed as:

$$J_{e,z}^{n+\frac{1}{2}}(i,j,k)=\frac{\Delta z}{2.R.\Delta x.\Delta y}\left(E_z^{n+1}(i,j,k)+E_z^{n}(i,j,k)\right) \qquad [3.120]$$

Substituting equation [3.41] into [3.2] will give:

$$E_z^{n+1}(i,j,k)=\left(\frac{1-\dfrac{dt.\Delta z}{2.R.\varepsilon.\Delta x.\Delta y}}{1+\dfrac{dt.\Delta z}{2.R.\varepsilon.\Delta x.\Delta y}}\right).E_z^{n}(i,j,k)+\left(\frac{\dfrac{dt}{\varepsilon}}{1+\dfrac{dt.\Delta z}{2.R.\varepsilon.\Delta x.\Delta y}}\right).\nabla\times H_z^{n+\frac{1}{2}}(i,j,k) \qquad [3.121]$$

3.4.7.2. *In the case of a capacitor*

The law of the current circulating in the capacitor according to the voltage is given by:

$$I = C\frac{\partial V}{\partial t} \qquad [3.122]$$

That is why the expression for the density of the current circulating in the capacitor oriented in the z axis at $n+1/2$ can be written as:

$$J_{e,z}^{n+\frac{1}{2}}(i,j,k) = \frac{C.\Delta z}{\Delta x.\Delta y.dt}\left(E_z^{n+1}(i,j,k) + E_z^{n}(i,j,k)\right) \qquad [3.123]$$

From [3.123], we get the FDTD equation that regulates the localized capacity Cx, deduced as follows:

$$E_z^{n+1}(i,j,k) = E_z^{n}(i,j,k) + \left(\frac{\frac{dt}{\varepsilon}}{1+\frac{C.\Delta z}{\varepsilon.\Delta x.\Delta y}}\right).\nabla\times H_z^{n+\frac{1}{2}}(i,j,k) \qquad [3.124]$$

3.4.7.3. *In the case of an inductor*

The current circulating through an inductor L is given by:

$$I = \frac{1}{L}\int_0^t V(\tau).d\tau \qquad [3.125]$$

In the case of an inductor oriented towards the z axis, the expression for the current density at $t+1/2$ is given by:

$$I_z^{n+\frac{1}{2}} = \frac{\Delta z.\Delta t}{L.\Delta x.\Delta y}\sum_{m=1}^{n} E_z^{m}(j,j,k) \qquad [3.126]$$

The FDTD equation that regulates the localized inductance L is then given by:

$$E_z^{n+1}(i,j,k) = E_z^n(i,j,k) + \frac{dt}{\varepsilon}.\nabla\times H_z^{n+\frac{1}{2}}(i,j,k) - \frac{\Delta z.dt^2}{\varepsilon.L.\Delta x.\Delta y}.\sum_{m=1}^{n} E_z^m(i,j,k) \quad [3.127]$$

3.4.8. *Interconnection simulations*

3.4.8.1. *For a microstrip line*

Let us consider a microstrip line. This microstrip consists of a conductive strip placed on a face of a dielectric material (resistive substrate), the other face of which constitutes a perfect conductor connecting the ground plane. The width of the microstrip line is $w = 1\ mm$ and the length is $L=20\ mm$. The relative dielectric permittivity of the resistive substrate is $\varepsilon_r = 4.0$, width $F= 10\ mm$ and thickness $h=1\ mm$ (Figure 3.14). The conductivity of the metallic strip and the ground place is $\sigma = 45.10^6\ S/m$. The surge impedance at the supply ports is $Z_0 = 50\ \Omega$, to avoid any impedance mismatch problems.

We applied a Gaussian potential difference at the input of the line, whose difference will vary with time between the two conductors.

An FDTD calculation code was created with MATLAB to simulate the microstrip line.

3.4.8.2. *For VIA interconnection*

In this section, we will run simulations on a microstrip line with a VIA interconnection (widely used to connect stacked circuits, in 3D stacking technology).

Simulating the model shown in Figure 3.55 is carried out using COMSOL, as it makes it possible to create a device and to highlight the main parameters.

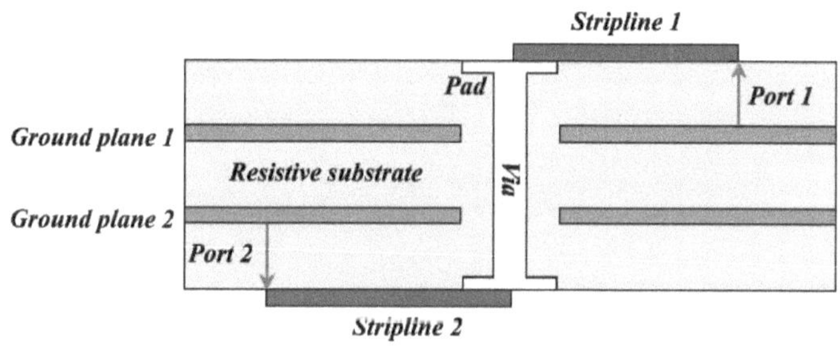

Figure 3.55. *Connection between two microstrip lines with VIA. For a color version of this figure, see www.iste.co.uk/gontrand/nanoelectronics.zip*

The model shown here is made of a VIA connected by two conductor sections and a silicon substrate; the electromagnetic signal is connected to the supply port with surge impedance $Z_0=50\ \Omega$. The parameters of the model studied are given in Table 3.1.

	Strip	VIA	Pad	Ground place	Substrate
Conductivity	45.10^6 S/m	45.10^6 S/m	45.10^6 S/m	45.10^6 S/m	10 S/m
Relative permittivity	1	1	1	1	11.7
Width	1 mm			0.02	0.02
Thickness	100 µm	5 mm	100 µm	100 µm	0.2 mm
Diameter		1 mm	1.5 mm		

Table 3.1. *Geometric and electrical parameters of the model studied*

The total electric field is shown in Figure 3.56 for *1 GHz* frequency. It is clear that the electric field is guided from the upper microstrip line to the lower microstrip line by the VIA with a loss in electrical energy.

Figure 3.20 shows the reflection factor S_{11} and the transmission factor S_{21} for a frequency range of *0* to *100 GHz*. It can be noted that the use of VIAs induces power dissipation (loss of gain) and several reflection peaks.

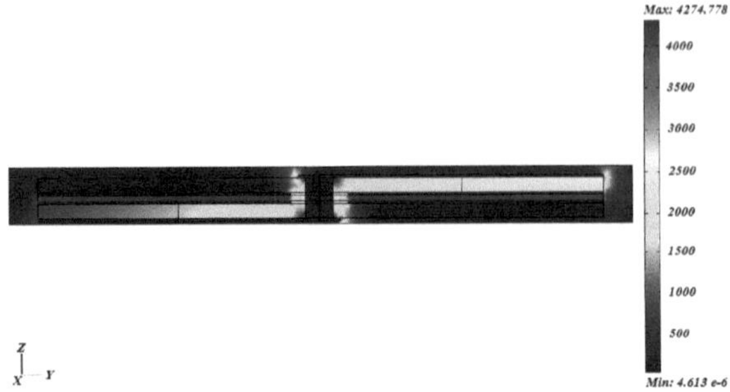

Figure 3.56. *Distribution of the total electric field in a VIA. For a color version of this figure, see www.iste.co.uk/gontrand/nanoelectronics.zip*

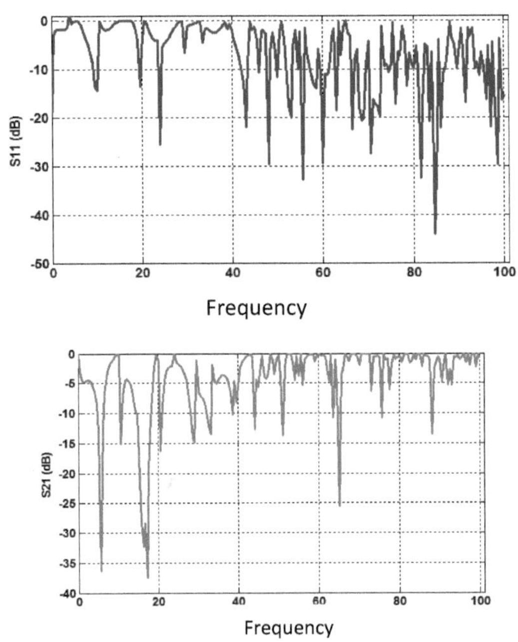

Figure 3.57. *Simulation of parameters S from 0 to 100 GHz using COMSOL. For a color version of this figure, see www.iste.co.uk/gontrand/nanoelectronics.zip*

The same simulation is run; however, in this case, the two conductor sections are perpendicular. Figure 3.58 shows the mapping of the total electric field for *1 GHz* frequency.

Parameters S of the perpendicular microstrip line structure are shown in Figure 3.22. From the results obtained, we see that there is energy loss when compared with the results obtained using a direct microstrip line structure.

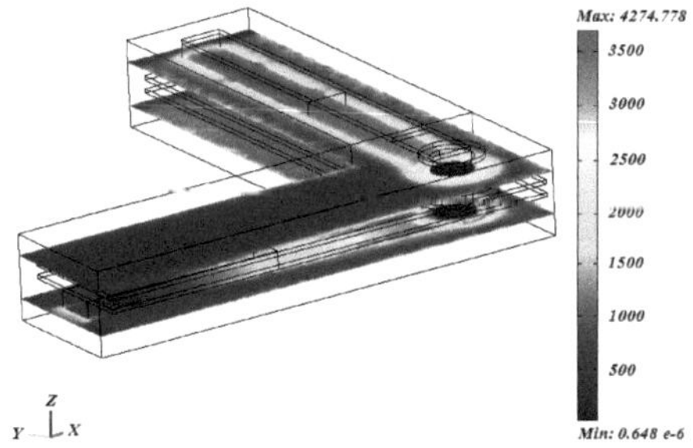

Figure 3.58. *Distribution of the total electric field in a perpendicular VIA. For a color version of this figure, see www.iste.co.uk/gontrand/nanoelectronics.zip*

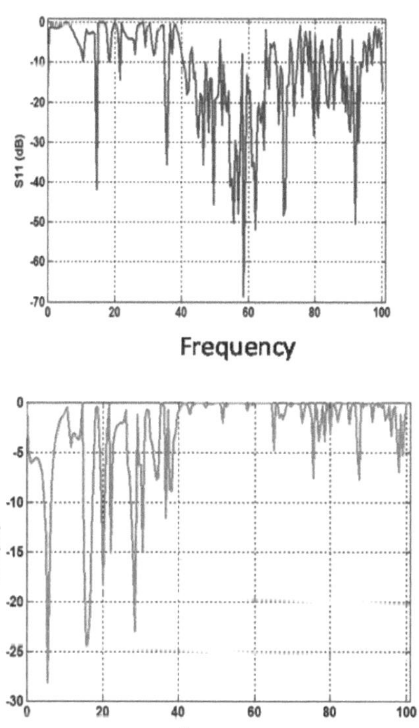

Figure 3.59. *Simulation of parameters S from 0 to 100 GHz using COMSOL. For a color version of this figure, see www.iste.co.uk/gontrand/nanoelectronics.zip*

Finally, we will run simulations of a U-shaped microstrip line with two VIAs. Figure 3.23 shows the mapping of the total electric field for *1 GHz* frequency. Parameters S of the structure are shown in Figure 3.61.

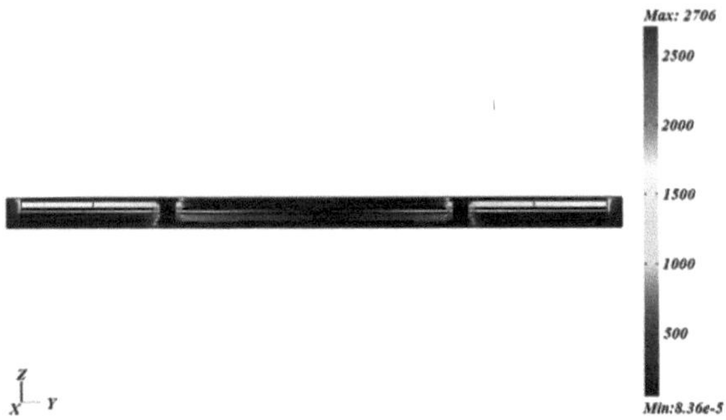

Figure 3.60. *Distribution of the total electric field of a U-shaped line. For a color version of this figure, see www.iste.co.uk/gontrand/nanoelectronics.zip*

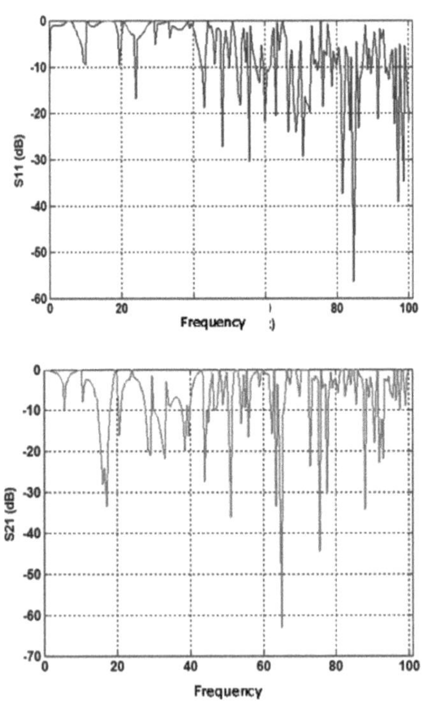

Figure 3.61. *Simulation of parameters S from 0 to 100 GHz using COMSOL. For a color version of this figure, see www.iste.co.uk/gontrand/nanoelectronics.zip*

3.4.8.3. *Radiofrequency and microwave circuits*

A simulation example is shown in Figure 3.62. In radiofrequency, the analysis of the MOS transistor requires that the effects of the interconnections are considered. Thus, the proposed circuit consists of a MOS transistor and connectors (microstrip lines), with holes and two resistors. The 3D microwave circuit is made via MMIC (monolithic microwave integrated circuit) technology; the thickness is 100 μm and the relative dielectric constant of the substrate is 3.9. The microwave circuit is designed as follows:

– the active device (MOS transistor) is made of two oriented current sources in the x and y directions;

– metal strips are used to interconnect the devices of the circuit; these interconnections are modeled like transmission lines of thickness 35 um;

– the source contact and load resistor are connected to the ground place using interconnection holes in the circuit;

– input and output ports for RF signals are connected in such a way as to have 50 Ω transmission lines;

– the excitation pulse is Gaussian, with a maximum frequency of 50 GHz.

LE-FDTD simulation was run using a uniform space grid: Dx = 0.189 mm, Dy = 0.2 mm and Az = 0.02 mm. In order to take into account the propagation along an infinite medium, the circuit was surrounded via UPML (Uniaxial Perfectly Matched Layer) boundary conditions: [L_i, J and C]. Perfect electric conductor (PEC) was applied in order to define metal surfaces and interconnection holes. The active circuit was polarized via a potential V_G= 0.8 V and V_D=1V. Mapping of the vertical device of the electric field is shown in Figure 3.63 (normal device of the electric field in comparison to the surface of the circuit), from which the non-uniform potential distribution along the conductive lines may be clearly deduced. Parameters S and phase analysis for nMOS transistor were obtained and are shown in Figure 3.62 for the entire range of interval frequency (0.1–50 GHz). The matrix of diffusion was extracted from the time responses for gate and drain contacts. To estimate parameters S for the microwave circuit, the output port 2 was open [X. Ye], whereas the Gaussian pulse was applied to port 1. The frequency spectrum of the signal was extracted via an FFT analysis from the transient time response. Figure 3.63 shows the transmission and reflection coefficients calculated at the two active port levels (S_{11} and S_{21}). To confirm these results, a simulation was run on the same structure using two commercial simulators (HFSS and ISE-TCAD). In this case, we extracted the matrix of the ISE device parameters S from the ISE-TCAD simulator, whereas the passive circuit was analyzed using the HFSS simulation software. Light differences between LE-FDTD and the finite differences method are *a priori* mostly due to the solution method and the absorbing boundary conditions.

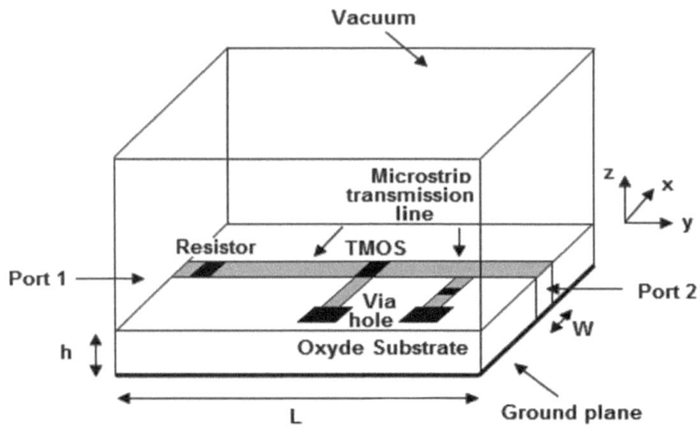

Figure 3.62. *Simulated structure*

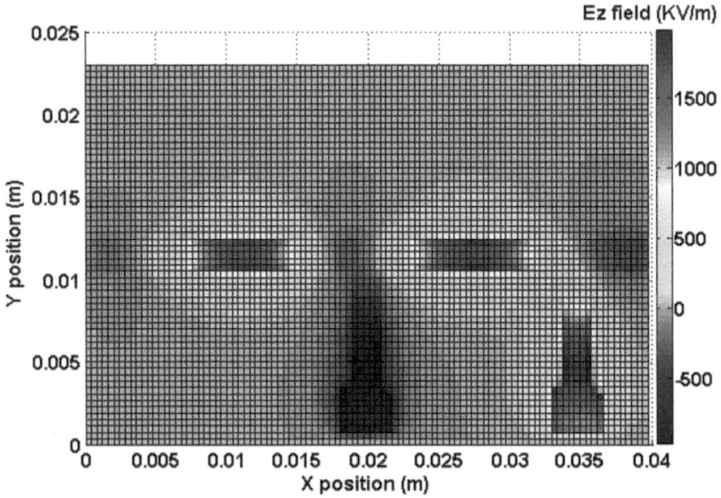

Figure 3.63. *Map of the electric field after 700 time steps. For a color version of this figure, see www.iste.co.uk/gontrand/nanoelectronics.zip*

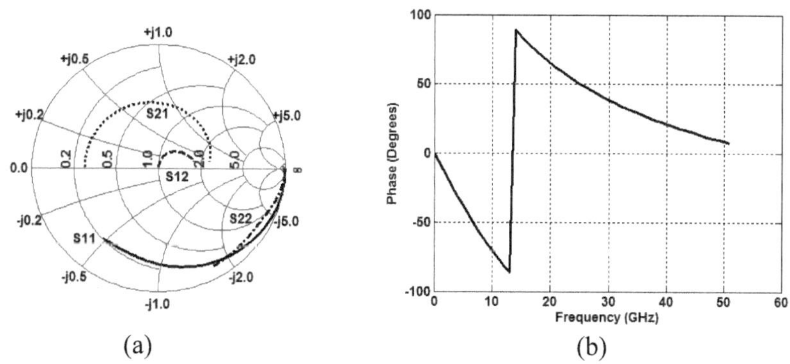

Figure 3.64. *Frequency analysis of an nMOS transistor from 0.1 to 50 GHz: a) parameters S (Smith chart); b) phase output analysis: drain contact*

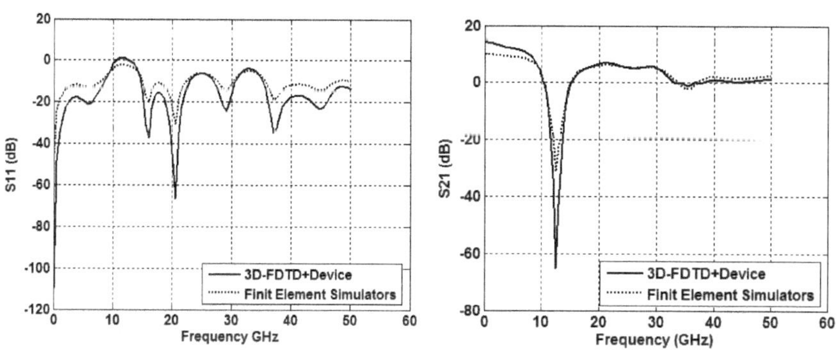

Figure 3.65. *Parameters S calculated for the radiofrequency circuit*

Presenting key submicron devices, such as TBH, DGMOSFET, MOS RF, and numerical methods (Wigner/Monte-Carlo, Poisson/Schrödinger, DDM/Maxwell) developed to establish electric characteristics, indicates future trends, leading to the use of the latest theoretical insights in nanodevices.

Simulation results, among others, show that the connections between devices, metal strips/RDL (Redistribution Layer), through holes, general electrical connectors, 3D, must be considered more and more in electromagnetic simulation in radiofrequency and microwave circuits.

3.5. Substrate in tridimensional circuits at micro- and nanoelectronic levels

3.5.1. *Introduction*

Here, we are concerned about 3D integration, modeling, from a physical point of view, in the electrical and temperature domains in transient regimes, applied to multilayer silicon substrates. This work is driven by the need to develop analytical solutions that are more manageable than numerical methods, such as finite differences or finite elements, of which there is a great demand in the industry.

First, we calculate the propagation impedance from any source. To do this, we have two mathematical tools: Green's kernel or the transmission-line matrix. The Poisson equation must be solved analytically. Cosine transforms, widely used to compress data and their variations, are used to accelerate calculation. All solutions to the system of equations are in a reciprocal system, the real values being extracted only at the end of calculation. We apply our algorithms using the MATLAB software. Thus, we can extract 3D impedances between two embedded or non-embedded contacts in the circuit of any form: physical or virtual. Then, we calculate the temperature for any source contact in the volume where hot spots can appear, induced by high-energy transient regimes, via the heat equation. We study our models via analytical and numerical methods, such as finite element simulation. They are completely compatible with software based on solutions to Kirchhoff equations, such as SPICE, and must allow for an in-depth study of electro-thermal impact of embedded contacts or surface-level contacts at any point in the circuit.

Nomenclature

a,b: chip size (m)

M,N: calculation points on the surface a*b

ρ: density of silicon ($kg.m^{-3}$)

C: isobaric heat capacity (J/(kg.K))

k: effective isotopic thermal conductivity (W/(m.K))

α ($= k / (\rho c)$): thermal diffusivity $\Omega.m^{-1}$

t_l: thickness of the layer l (m)

V, ϕ: potentials (V)

σ: conductivity of silicon (S/m)

λ (or μ or γ): intrinsic value (m^{-1})

ε: permittivity of silicon (F/m)

T: temperature (°C)

R_{FCS}: spreading impedance coefficients

ω: angular frequency (s^{-1})

Electro-thermal coupling in the substrate of integrated circuits is a significant consideration in the design. Of late, micro (nano) technology and the development of semiconductor technology has allowed designers to integrate multiple systems in a single chip, not only in 2D (on the surface), but also in 3D. This design technology also reduces the integration cost; it seems to be a good competition for the Moore law (more than Moore against more Moore). However, there is still work to be done. In 2010, the burning question was, "why 3D?" Today, the questions are, "when and how 3D?" Stacking sublayers of 3D circuits is considered a way to surmount conventional 2D-IC locks using 3D interconnections, inter-strata VIAs for the transmission of signals.

The type of silicon substrate in which 3D interconnections are fabricated is an important parameter and has a strong impact on electrical behavior, also depending on the application. Two types of silicon substrates P are used typically in the CMOS/BiCMOS process: one uniform and lightly doped (> 1 Ω.cm) and one strongly doped (< 1 Ω.cm) with (1 Ω.cm) a fine, lightly doped epitaxial layer. Other semiconductor substrates are used in micro/nano technologies and may include 3D interconnections (SOI technology, MEMS technology, sensor technology, etc.). Here, we try to start a summary, via transient electro-thermal studies in circuits in 3D. First, we solve the Poisson equation using Green's kernel or TLM, based on the reflection–transmission of the voltage and current.

Then, we work on solving the 3D non-stationary heat equation; as much as possible, we use algorithms developed for the electrical model. In conclusion, we prove that coupled electric and thermal diffusion impedances can be manipulated, and propose to develop this concept of diffusion/propagation impedance for 3D circuits.

3.5.2. *Mathematical tools*

Green's function kernel calculations, as opposed to numerical methods, are applied to stacked homogeneous layers constituting the substrate to extract the model of this substrate; the speed of resolution of Green's function method is very rapid. Some basic reminders and the concept are presented. The use of discrete cosine transforms (DCT), using an FFT twice, for instance – FFT: rapid Fourier transform – accelerates the speed of the calculation. Then, a full model, which can be applied to a substrate with deep contacts (or VIA), is presented; this deals with contacts found in different layers. To verify, we compare with a finite element method (FEM) [COM].

3.5.2.1. *Green's kernel*

High-density integration and high frequency of the system make coupling of the noise of the substrate an important point in the design, due to its impact on the performance of integrated circuits. The objective of this 3D analysis of the substrate is first to efficiently extract impedance parameters Z between two contacts (electrodes, VIA), which are located on or in the silicon substrate. An efficient tool for extracting impedance for contacts may be to help the designer to accelerate designing as well as optimize final mask design. As with planar technology, 3D interconnections may be modeled using an RLCG divider: (G // C)/(R series L), this equivalent electrical model being in Π or T network. Often, a simple compact model is represented as a simple node. However, this hypothesis is viable for a substrate that is very conductive in low and medium frequencies, and does not work in the case of multilayer substrates. This is the reason why, in this section, we propose a method of extracting the substrate that is based on transmission-line matrix (TLM) or Green's methods [GHA 95, NIK 09].

3.5.2.2. *Substrate analysis*

The best-known groundbreaking reference is Gharpurey [GHA 95]: a layered 3D substrate with two surface contacts. In general, parameters Z may be defined as:

$$Z_{mn} = \frac{V_m}{I_n}\bigg|_{I_{k\neq n} = 0} \qquad [3.128]$$

Practically, by injecting a single-channel current as an excitation source I in a point n and a single excitation current source -I (equivalent to the current tank I) to a point m and calculating the potential at point m of the sensor, we obtain the impedance Z_{mn} between contacts m and n straightaway. If we can obtain the resulting potential distribution in the substrate, caused by the single-channel current, we can obtain parameters Z by [3.128]. In quasi-static conditions, the substrate potential satisfies the Laplace equation $\nabla^2\varphi(x,y,z) = 0$.

We can rewrite this equation as:

$$\left(\frac{\partial^2}{\partial x^2} + \frac{\partial^2}{\partial y^2} + \frac{\partial^2}{\partial z^2}\right)\varphi(x,y,z) = 0 \qquad [3.129]$$

ϕ= X(x,x′). Y(y,y′). Z(z,z′)

we can write:

$$Y \cdot Z \cdot \frac{d^2X}{dx^2} + X \cdot Z \cdot \frac{d^2Y}{dy^2} + X \cdot Y \cdot \frac{d^2Z}{dz^2} = -\frac{\ddot{a}(x-x')\cdot\ddot{a}(y-y')\cdot\ddot{a}(z-z')}{\mathring{a}_N} \quad [3.130]$$

Hence:

$$Y \cdot Z \cdot \frac{d^2X}{dx^2} + X \cdot Z \cdot \frac{d^2Y}{dy^2} + X \cdot Y \cdot \frac{d^2Z}{dz^2} = -\frac{\delta(x-x')\cdot\delta(y-y')\cdot\delta(z-z')}{\varepsilon} \quad [3.131]$$

and thus:

$$\frac{d^2X}{dx^2} + \lambda^2 X = \frac{d^2Y}{dy^2} + \mu^2 Y = \frac{d^2Z}{dz^2} - \lambda^2 + \mu^2 Z = 0 \quad [3.132]$$

where intrinsic values λ and μ are homogeneous over a wavelength (m^{-1}) for $x \neq x'$ $y \neq y'$ $z \neq z'$.

ϕ ι: Green's kernel l:

$$X = \cos\left(\frac{m\cdot\pi\cdot x}{a}\right); Y = \cos\left(\frac{n\cdot\pi\cdot y}{b}\right); \text{where m and n} \in [0,\infty) \quad [3.133]$$

$$\sum_{m=0}^{\infty}\sum_{n=0}^{\infty} \cos\left(\frac{m\cdot\pi\cdot x}{a}\right)\cdot\cos\left(\frac{n\cdot\pi\cdot y}{b}\right)\cdot\left[\frac{d^2Z}{dz^2} - \left(\left(\frac{m\cdot\pi\cdot x}{a}\right)^2 + \left(\frac{n\cdot\pi\cdot y}{b}\right)^2\right)\cdot Z\right] = -\frac{\delta(x-x')\cdot\delta(y-y')\cdot\delta(z-z')}{\varepsilon_N} \quad [3.134]$$

By definition:

$$Z(z,z') = Z'(z,z')\cdot\cos\left(\frac{m\cdot\pi\cdot x'}{a}\right)\cdot\cos\left(\frac{n\cdot\pi\cdot y'}{b}\right) \quad [3.135]$$

where:

$$\frac{ab}{4}\cdot\left[\frac{d^2Z}{dz^2} - \gamma_{mn}{}^2\cdot Z\right] = -\frac{\delta(z-z')}{\varepsilon_N} \text{ with } \gamma_{mn} = \sqrt{\left(\frac{m\cdot\eth}{a}\right)^2 + \left(\frac{n\cdot\eth}{b}\right)^2} = \sqrt{(\lambda)^2 + (\mu)^2} \quad [3.136]$$

For $z \neq z'$, $\delta(z-z') = 0$. Equation [3.132] has the well-known general solution:

$$Z' = A\cdot e^{-\tilde{a}_{mn}\cdot(d+z)} + B\cdot e^{\tilde{a}_{mn}\cdot(d+z)} \quad [3.137]$$

This equation calls for a transmitted wave and a reflected wave. A general solution for m > 0 and n > 0 of Green's kernel is given by:

$$\emptyset\,(x, y, x', y', z, z') = \sum_{m=1}^{\infty}\sum_{n=1}^{\infty}\left[\frac{2\cdot\left(A_N e^{-\gamma_{mn}(d+Z_l)}+B_N e^{\gamma_{mn}(d+Z_l)}\right)\left(e^{-\gamma_{mn}Z_h}+e^{\gamma_{mn}Z_h}\right)}{ab\varepsilon_N\gamma_{mn}\left(B_N e^{\gamma d}-A_N e^{-\gamma d}\right)}\right]\cos\left(\frac{m\pi x}{a}\right)\cos\left(\frac{n\pi y}{b}\right) \quad [3.138]$$

where $Z_{l(ow)} = min(z, z')\ and\ Z_{h(igh)} = \max(z, z')$

For example, let us consider a case where the load point and the observation point are in the same layer on the surface and z = z' = 0. Green's function becomes:

$$\emptyset(x, y; x', y')_{z=z'=0}\ (G_0)_{z=z'=0} + \left(\sum_{m=0}^{\infty}\sum_{n=0}^{\infty} f_{mn}C_{mn}\cos\left(\frac{m\pi x}{a}\right)\cos\left(\frac{m\pi x'}{a}\right)\cos\left(\frac{n\pi y}{b}\right)\cos\left(\frac{n\pi y'}{b}\right)\right) \quad [3.139]$$

where C_{mn}= 0 for (m=n=0), C_{mn}=2 for m=0 or n=0 but m ≠ n and C_{mn}= 4 for all other m and n (m > 0 and n > 0).

The f_{mn} function is given by:

$$f_{mn} = \frac{1}{ab\tilde{a}\mathring{a}_N}\frac{\hat{a}_N\, tanh(\tilde{a}_{mn}d)+\tilde{A}_N}{\hat{a}_N+\tilde{A}_N\, tanh(\tilde{a}_{mn}d)} \quad [3.140]$$

β_N and Γ_N can be calculated recursively:

$$\begin{bmatrix}\beta_k\\ \Gamma_k\end{bmatrix} = \frac{1}{2}\cdot\begin{bmatrix}1+\frac{\varepsilon_{k-1}}{\varepsilon_k} & (1-\frac{\varepsilon_{k-1}}{\varepsilon_k})e^{2\theta_k}\\ (1-\frac{\varepsilon_{k-1}}{\varepsilon_k})e^{-2\theta_k} & 1+\frac{\varepsilon_{k-1}}{\varepsilon_k}\end{bmatrix}\begin{bmatrix}\beta_{k-1}\\ \Gamma_{k-1}\end{bmatrix} \quad [3.141]$$

where $\theta_k = \gamma_{mn}(d - d_k), \beta_0 = 1, \Gamma_0 = 1$

After deriving a formal mathematical solution to equation [3.132] with related boundary conditions, we will discuss a more physical (and handier) approach to the problem, using the transmission-line method, with a few recapitulatory statements.

Note that in our entire algorithm, we are working in the reciprocal space right from the very beginning. After a spatial two-dimensional Fourier transform in the x and y directions, equation [3.132] should be:

$$\frac{\partial^2\Phi}{\partial z^2} - k_x^{\,2}\Phi - k_y^{\,2}\Phi = 0 \quad \text{– cf. eq. [3.132]} \quad [3.142]$$

where Φ is the potential and k_x ky are space–frequency variables (wave vectors) (m^{-1}).

Current density $J_Z = \sigma^* E_Z$ (σ* is the complex conductivity), and $E_Z = -\frac{\partial \varphi}{\partial z}$; thus:

$$J_z = -\sigma^* \frac{\partial \varphi}{\partial z} \qquad [3.143]$$

In space,

$$J_z\left(k_x, k_y, z\right) = -(\sigma + j\omega\varepsilon)\frac{\partial \Phi\left(k_x, k_y, z\right)}{\partial z} \qquad [3.144]$$

Thus:

$$\frac{\partial J_z}{\partial z} = -(\sigma + j\omega\varepsilon)\frac{\partial^2 \Phi}{\partial z^2} \qquad [3.145]$$

From [3.142] and [3.145], we have:

$$\frac{\partial J_z}{\partial z} = -(\sigma + j\omega\varepsilon)\left(k_x^{\ 2} + k_y^{\ 2}\right)\Phi \qquad [3.146]$$

Modifying [3.142] and [3.145], we finally obtain:

$$\begin{cases} \dfrac{\partial \Phi}{\partial z} = -\dfrac{J_z}{(\sigma + j\omega\varepsilon)} \\ \dfrac{\partial J_z}{\partial z} = -(\sigma + j\omega\varepsilon)\left(k_x^{\ 2} + k_y^{\ 2}\right)\Phi \end{cases} \qquad [3.147]$$

These equations show the relation between current density in direction z and the potential distribution in the space–frequency domain.

3.5.3. *Analogy with transmission lines for multilayer substrates*

In its simplest form, a transmission line is a pair of conductors connecting two electric systems (source and load, for instance), with a two-way path; in the case of a floating return path, a third (or more) conductor becomes the ground. For microwaves, these are waveguides. In our case, electromagnetic wave propagation and interferences through the silicon substrate are among the most crippling obstacles in the existing trend of integration in microelectronics. TLM can be seen as a physical interpretation of the above-mentioned mathematical developments. The main strength of this method is that it seems to be meant for embedded contacts, irrespective of their quantity and form.

A plane wave is a wave whose incidence plane is parallel to the plane <z', z>. Let us consider the general case where the impedance of the line is not the same as that of load. A wavefront A hits load Z_L: some energy is absorbed by Z_L and the remaining energy is reflected; in this case, voltage and current wavefronts are not in phase. This reflected wave may meet another incident wavefront. The direction of the current flux depends on the polarity of the related wave:

$$Z = \frac{V_{total}}{i_{total}} = \frac{V_F \pm V_R}{i_F \mp i_R} \quad [3.148]$$

The general solution for the voltage and current is:

$$V(z) = V_f e^{\tilde{a}z} + V_r e^{-\tilde{a}z} \quad [3.149a]$$

$$I(z) = Y_c (V_F e^{-\tilde{a}z} - V_r e^{\tilde{a}z}) \quad [3.149b]$$

$\gamma = \alpha + \varphi\beta$; α: attenuation, $\beta = 360°/\lambda_{si}$.

$$\begin{bmatrix} V_1 \\ I_1 \end{bmatrix} = \begin{bmatrix} \cosh \alpha L & Z_c \sinh \alpha L \\ Y_c \sinh \alpha L & \cosh \alpha L \end{bmatrix} \begin{bmatrix} V_2 \\ -I_2 \end{bmatrix} \quad [3.150]$$

From the output branch $V_2 = -Z_L I_2$, we have:

$$\begin{bmatrix} V_1 \\ I_1 \end{bmatrix} = -I_2 \begin{bmatrix} Z_L \cosh \alpha L & Z_c \sinh \alpha L \\ Z_L Y_c \sinh \alpha L & \cosh \alpha L \end{bmatrix} \begin{bmatrix} 1 \\ 1 \end{bmatrix} \quad [3.151]$$

Z_L is the load impedance and L the load distance; we can extract the input impedance, Z_l (= V_1/I_1), a function of the load impedance Z_{l+1} (l: layer):

$$Z_l = Z_c \frac{Z_{l+1} + Z_c \tanh \alpha L}{Z_c + Z_{l+1} \tanh \alpha L} \quad [3.152]$$

where $Zc = \beta_N$ and $Z_L = \Gamma_N$ [3.141].

In fact, starting from equation [3.140] and taking boundary conditions into account – continuity of potential and discontinuity of an electric field, if there is a surface load at the interface of the concerned layer – it is easy to modify this iterative solution (versus substrate depth or layers) of equation [3.140]. In fact, these calculations are carried out in one go for each frequency.

In our simulator built via MATLAB [MAT], we use a matrix formalism, propagating impedance Z and the current transmittance factor from layer 1 to the adjacent layer (see equation [3.140] or equation [3.151]), from the layer k:

$$\begin{bmatrix} Z_{k,l+1} \\ T_{k,l+1} \end{bmatrix} = \begin{bmatrix} A_k & -B_k \\ C_k & -D_k \end{bmatrix} \begin{bmatrix} Z_{k,l} \\ T_{k,,l} \end{bmatrix} \quad [3.153]$$

T shows the branching off of current (between two parallel regions), calculations in a reciprocal space. For example, potential is derived from the product (Kronecker) of (sub-)matrices V using a DCT (Direct Cosines Transform) of the matrix of the injected current I; two consecutive FFTs will help get this DCT.

In our algorithm, contact voltage is the average voltage of discrete contact elements, calculated using the Millman theorem, such that:

$$V = \frac{\sum_{n=1}^{N} j\omega c \varphi_n}{\sum_{n=1}^{N} j\omega c} = \frac{1}{N}\sum_{n=1}^{N} \varphi_n \quad [3.154]$$

where φ_n is the voltage of the substrate of a node and N is the number of sub-zones of contact. Thus, in this new algorithm, the voltage of the contact is calculated by taking the average of potentials, <Va>, in the region of the substrate under contact with the injection.

3.5.4. *Green/TLM results*

In this section, a typical 0.35 μm BiCMOS technology is studied for verification purposes; in Figure 3.66, a particular region (P +/P) of this process is shown. We approach the actual profile by stacked layers of uniform thickness (taken *ex abrupto*) which are uniformly doped (Figure 3.66(b)).

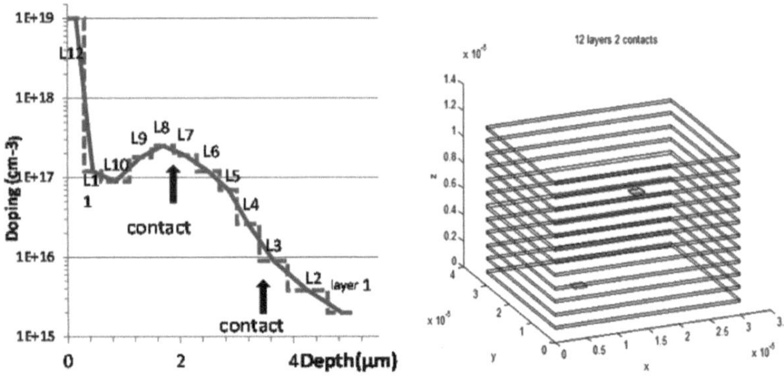

(a) (b)

Figure 3.66. *a) Profile; p+/p– region; BiCMOS technology: 0.35 µm; b) three schema of a 3D structure. For a color version of this figure, see www.iste.co.uk/gontrand/nanoelectronics.zip*

First, after numerical experiments using COMSOL [COM], a multiphysics (see electric, thermal, mechanical) simulator was used; this simulator is also used for full-wave electromagnetic analysis, going beyond quasi-stationary cases. It primarily uses Galerkine-like algorithms. In general, these 3D simulations can last anywhere from a few minutes to several hours.

3.5.4.1. *Embedded contacts*

Let us take a more general case where real and virtual contacts are completely embedded. Then, the substrate can be considered a connection of two parallel sections (branching off of current).

Now, it seems easy to calculate any transfer impedance by TLM. The chip surface is 30 µm * 30 µm, and M = N = 300; contact surface is 20 points into 20 points (calculations points, M * N, are equally distributed). In Figure 3.67, we present an impedance module between contacts Ci and Cj, at interface L3/L4 and L7/L8. Capacitance and resistance are extracted. It is typical of a Lorentzian function ($Z_{RC} \equiv R$ // C, and RC = $\rho\varepsilon$); TLM/Green and COMSOL match.

From the above results, we see that TLM/Green and COMSOL match, while the analytical method takes very little time (approximately 1/50, even 1/100).

The absence of uniform potential at the contacts level is caused due to the injection of a uniform current density instead of uniform voltage.

Simulation results obtained using Green/TLM from MATLAB are practically the same, compared with COMSOL. In fact, with MATLAB, we have the choice of using layers like perfect parallelepiped (layers with uniform thickness); in this case, equation [3.151] is an accurate analytical solution. Otherwise, with refined meshes relating to FEM methods, there is a single convergence of solution. If we design layers with a certain variation in thickness, the two methods will not give the same solutions, depending on the variation, because equation [3.151] is primarily a very accurate solution.

We are quite aware that quasi-electrostatic modeling cannot work perfectly at 10 THz (vs. experiments), which involves a wavelength of 10 µm. This wavelength is of the same magnitude as the contact-to-contact distance (3 µm) taken in Figure 3.66.

We have not explicitly taken permeability into account. It is not realistic to go beyond 200 GHz in this analysis (which is not too bad given current applications); but we think it will be interesting to make a comparison with the finite elements method to test the robustness of this algorithm.

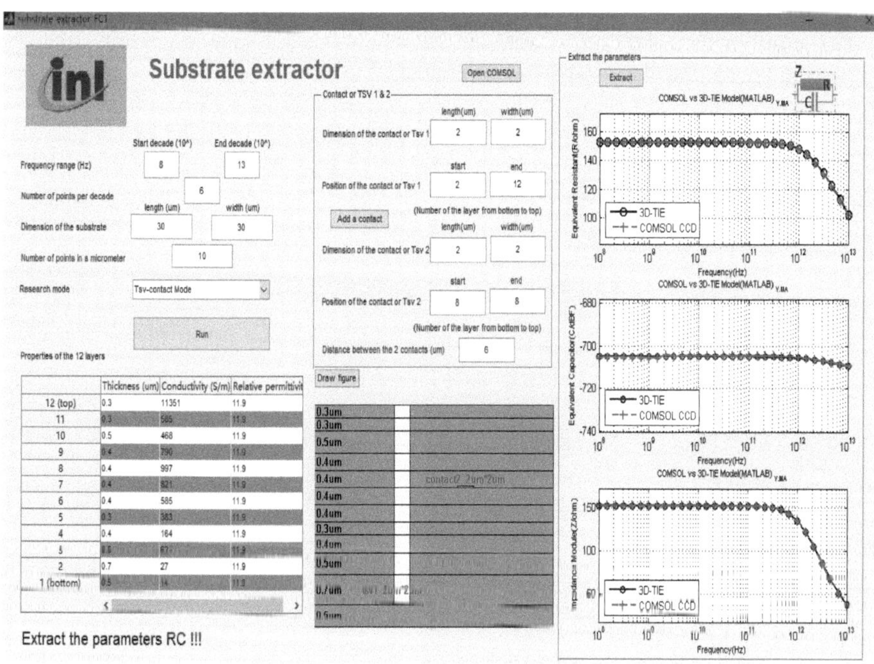

Figure 3.67. *Impedance module Z17, and extraction of capacitance and resistance C17 and R17. For a color version of this figure, see www.iste.co.uk/gontrand/nanoelectronics.zip*

3.5.5. *Heat equation*

Here, we deal with variable temperatures due to one or more heat sources, in this case using the superposition principle.

A diffusion resistance/thermal propagation – always in the reciprocal domain – can be extracted from the discrete heat source; typical applications are the cooling of localized zones called hot spots, at the packaging or chip level.

We take a transient three-dimensional heat equation in a finite, non-homogeneous medium.

We can write the heat expression (which is also the diffusion equation) as:

$$\frac{d^2T}{dx^2}+\frac{d^2T}{dy^2}+\frac{d^2T}{dz^2}-\frac{1}{\alpha}\frac{dT}{dt}=0 \qquad [3.155]$$

where $\alpha = k/(\rho c)$ is the thermal diffusivity, k the isotropic thermal conductivity and C the thermal isobaric mass.

For equilibrium, the heat equation becomes the Laplace equation, like the Poisson equation does in the absence of electrical load.

As we can see, there is an analogy between equations [3.131] and [3.155].

If we assume that T= X(x,x′). Y(y,y′). Z(z,z′) τ(t,t′), we can extend equations [3.131] and [3.155], such that:

$$\frac{1}{X}\cdot\frac{d^2X}{dx^2}+\frac{1}{Y}\cdot\frac{d^2Y}{dy^2}+\frac{1}{Z}\cdot\frac{d^2Z}{dz^2}-\frac{1}{\alpha}\cdot\frac{d/}{dt}=-A^2\cdot\frac{\delta(x-x')\cdot\delta(y-y')\cdot\delta(z-z')}{\alpha} \qquad [3.156]$$

Thus:

$$\frac{d^2X}{dx^2}+\lambda^2X= \quad \frac{d^2Y}{dy^2}+\mu^2Y= \quad \frac{d^2Z}{dz^2}-\gamma^2Z= \quad \frac{d\tau}{dt}-\alpha(\lambda^2+\mu^2-\gamma^2)\,\tau=0 \qquad [3.157]$$

or:

$$\lambda^2+\mu^2-\gamma^2+\frac{A^2}{\alpha}=0 \qquad [3.158]$$

This last equation in [3.157] has an evanescent solution according to time: for layer i, we can write:

$$\tau_i = \tau_{i0}..\exp(-(+\lambda_{ix}^2 + \mu_{iy}^2 - \gamma_{iz}^2).\alpha.t) \quad [3.159]$$

Still in the reciprocal domain (thus, we can make use of compression, which is inherent to DCT), equation [3.157] becomes:

$$\frac{d^2X}{dx^2} + \frac{d^2Y}{dy^2} + \frac{d^2Z}{dz^2} - \frac{1}{\alpha}.\frac{d\tau}{dt} = \lambda^2 X \ \mu^2 Y + \gamma^2 Z - \left(\frac{j}{\alpha}\right)\omega = 0 \ (j^2 = -1) \quad [3.160]$$

We can define diffusion/propagation Fourier coefficients of the impedance function (transcendental) as follows:

$$R_{FCS\,i} = A_i/B_i \quad [3.161]$$

where A_i and B_i are Fourier coefficients of the general solution for temperature in the z direction perpendicular to the stacked layers.

$$FCSi = Ai.\cosh(\gamma z) + Bi.\sinh(\gamma z) \quad [3.162]$$

Boundary conditions

Continuity of thermal conductivity at the interfaces of the layers leads to:

$$k_i \frac{dT_{1i}}{dz,i} = k_{i+1}\frac{dT_{i+1}}{dz_{i+1}} \text{ (Neumann condition)} \quad [3.163]$$

At the interface, we also have:

$$X_i(t) = X_{i+1}(t),\ Y_i(t) = Y_{i+1}(t) \text{ (Dirichlet condition)} \quad [3.164]$$

$$\text{and } k_i \frac{T_{i,z}}{T_{i+1,z}} = k_{i+1}\frac{R_{FCS\,i,z}}{R_{FCS\,i+1,z}} \quad [3.165]$$

see the analogy with the induction vector; at an interface without fixed charges:

$$\varepsilon_i E_i = \varepsilon_{i+1} E_{i+1} \text{ (E: Electrical field)} \quad [3.166]$$

After analyzing equations [3.163] and [3.165], we obtain:

$$B_{iz}.R_{FCS\,i+1,z}.Y_{i(y)}.e^{-(-\gamma_{iz}^2 + \mu_{iy}^2 + \lambda_{ix}^2)t} =$$
$$-R_{FCS\,i\,z}.Y_{i+1(y)}.e^{-(-\gamma_{i+1,z}^2 + \mu_{i+1,y}^2 + \lambda_{i+1,x}^2)} \quad [3.167a]$$

$$k_{i,z}.R_{FCS\,i+1,z}.Y_{i(y)}.e^{-(-\gamma_{iz}^2 + \mu_{iy}^2 + \lambda_{ix}^2)t} =$$
$$k_{i+1,z}.B_{iz}.R_{FCS\,i,z}.Y_{i+1(y)}.e^{(-\gamma_{i+1,z}^2 + \mu_{i+1,y}^2 + \lambda_{i+1,x}^2)t} \quad [3.167b]$$

$$\gamma_{i+1\ z} = \left[\frac{\alpha_i}{\alpha_{i+1}} \cdot \gamma_{iz}^2 + \left(\frac{\alpha_i}{\alpha_{i+1}} - 1\right) \cdot (\lambda_{ix}^2 + \mu_{iy}^2)\right]^{\frac{1}{2}} \quad [3.168]$$

In fine, we get:

$$A_{iz} = B_{iz} R_{FCD\ i\ z} \cdot (\gamma_{i\ z}) \quad [3.169]$$

$$R_{FCD\ i\ z}(\gamma_n) = \frac{k_{i+1} R_{FCD\ i+1,z}(\gamma_n) + k_i . tanh(\gamma_{i\ z} . t_i)}{k_i + R_{FCD\ i+1,z}(\gamma_n) . tanh(\gamma_{i\ z} . t_i)} \quad [3.170]$$

(see analogy with equation [3.26]).

The impedance related to the development of the rapport of Fourier coefficients with the temperature solution is given in Figure 3.68.

Finally, the real general solution of equation [3.155] is as follows:

$$(T - T_\infty)_{(x,y,x',y',z,z')} =$$

$$T_0 . \exp(-(\lambda_{ix}^2 + \mu_{iy}^2 - \gamma_{iz}^2) . \alpha t) .$$

$$s \sum_{m=1}^{\infty} \sum_{m=1}^{\infty} \left[\frac{2\left(A_N e^{-\gamma_{mn}(d+z_l)} + B_N e^{\gamma_{mn}(d+z_l)}\right)\left(e^{-\gamma_{mn} z_h} + e^{\gamma_{mn} z_h}\right)}{ab \sum_N \gamma_{mn}\left(B_N e^{\gamma_{mn} d} - A_N e^{-\gamma_{mn} d}\right)}\right] \cos\left(\frac{m\ x}{a}\right) \cos\left(\frac{n\ y}{b}\right) \quad [3.171]$$

The first results are given in Figure 3.69 for temperature, resulting from both FEM methods and Green's algorithms, where the heat source profile applied to one or two contacts is shown in Figure 3.30 and the temperature map in Figure 3.40; FEM results are to the left and Green results to the right. We think that this type of modeling can be very useful to the industry concerned. For example, with electric 3D simulations, we can search for hot spot(s); using the thermal approach, we will have to work on the place and route of coolers (heat pipes, etc.), via simulations run on 3D heat propagation.

A minimal, electro-thermal coupling can be carried out using the simple linear formula:

$$R = R0\ (1 + \alpha \Delta T) \quad [3.172]$$

Analytical solutions are thus obtained for evaluating the transfer of electrical impedance and temperature distribution in the multilayer substrate. Green's kernel and transmission-line matrix, which give accurate analytical solutions, are the common approaches. One aspect of the above modeling is the frequent use of the analogy between the two domains: electrical and thermal. Efficient and perhaps elegant techniques for modeling the substrate in 3D circuits (via a frequency wobulation, followed by extracting parameters: capacitance, resistance, electrical, thermal) are presented and programmed. Using the FFT or DCT accelerates the calculation; this technique can make the optimization of the layout (mask design) of circuits very easy, to minimize the coupling effects on the substrate. Moreover, older models were often limited in the sense of surface contacts.

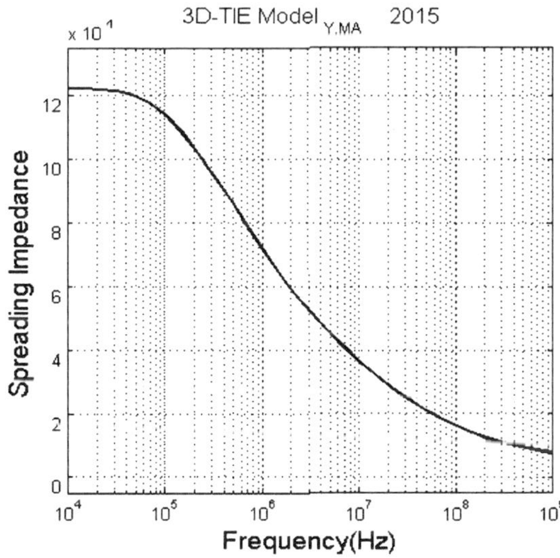

Figure 3.68. *Fourier coefficients of the temperature – spreading impedance*

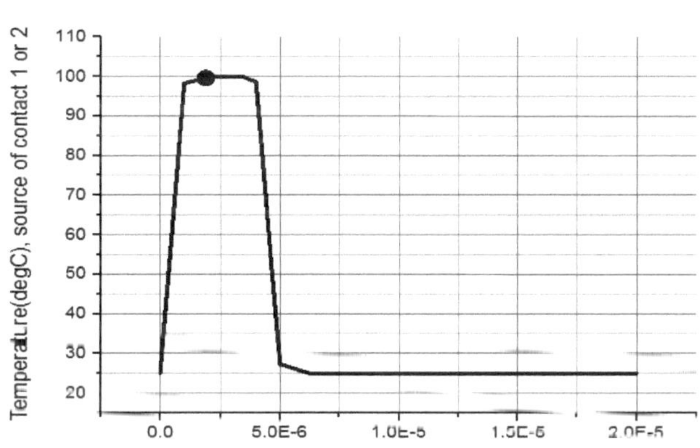

Figure 3.69. *Profile of the temperature source; measuring point at 2.2 μs*

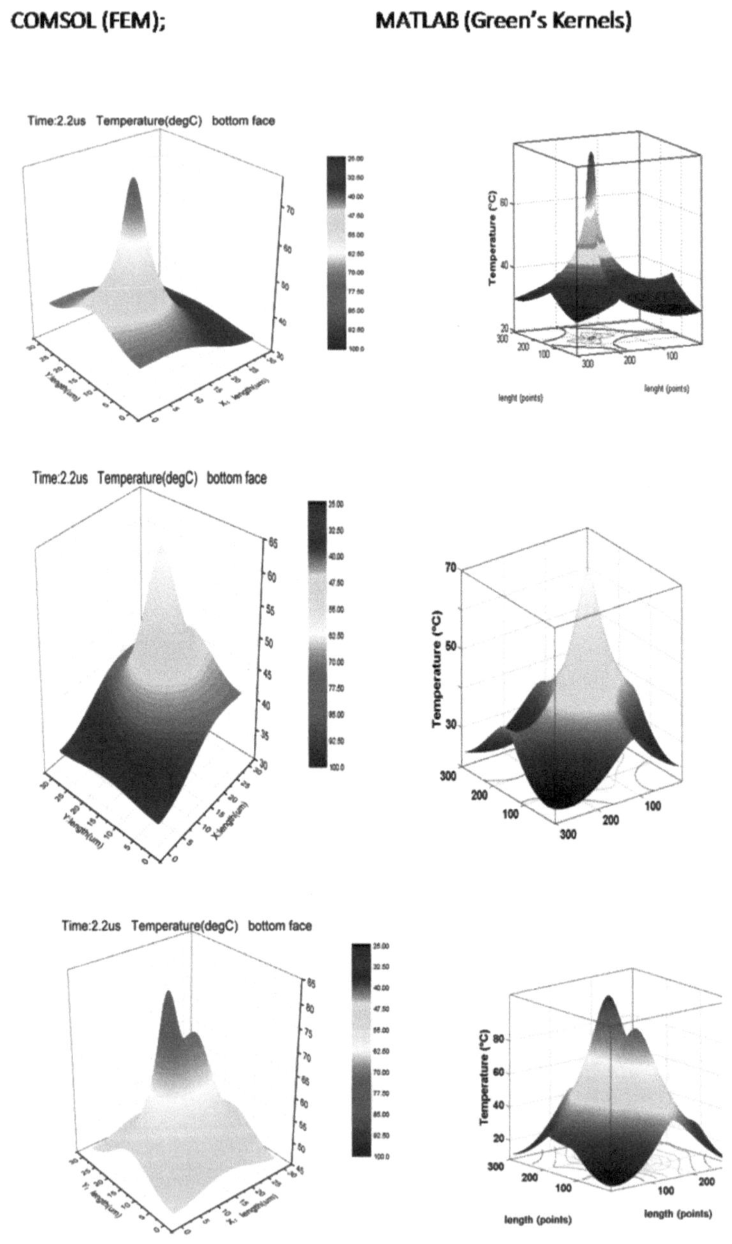

Figure 3.70. *Temperature mapping; left: FEM results, right: Green/TLM results. t =2.3 μs (Figure 3.69). For a color version of this figure, see www.iste.co.uk/gontrand/nanoelectronics.zip*

3.6. Conclusion

The continuous miniaturization of transistors over the last 40 years has led microelectronics to the nanometer scale, giving birth to nanotechnology! Comparing it with the human brain, we have learnt that the brain has way more elements – in the order of 1,000 billion – than we know how to assemble on an integrated circuit. Secondly, the brain develops in volume, whereas integrated circuits are primarily flat. Thus, there is a research dimension to be gained in this new field. Difficult scientific and technological challenges justify research programs that are more internationalized, with increasingly significant economic stakes.

Furthermore, starting from the key model for the current study of electrical behavior of devices, called the drift diffusion model (DDM), which is adapted to microelectronics, we go back to general equations, called master equations, such as Boltzmann and Wigner equations.

For nanotechnology, models that take into account more intransient phenomena, for example carrier overspeed, non-homogeneity of temperature in devices, as well as quantum phenomena, should be introduced.

3.7. Appendix

The program that we created is a numerical 2D simulation of a diode in Maple, a simulation that is governed by the drift diffusion model.

Here, we propose a simulation method that is obtained notably thanks to solving Poisson equations with the help of Newton algorithm as well as the use of the Scharfetter and Gummel algorithm for electron and hole transport.

This document briefly explains the working of this program, which models a P–N junction, and also gives an overview of the program's theoretical foundations.

Modeling the P–N (diode) junction was achieved through reverse engineering, using the Scharfetter–Gummel algorithm. This algorithm was made in 1969 by Hermann Gummel and Donald Lee Scharfetter (1934–2000) and makes it possible to numerically solve equations that govern potential and load in the junction.

3.7.1. *P–N junction: working; recap*

A diode or P–N junction is an electronic device made by juxtaposing two regions of the same monocrystalline semiconductor that have been doped differently.

The operation principal of the junction can be explained from a microscopic point of view.

We know that when two atoms are far apart, their electronic states are represented by atomic orbitals. When atoms are near, their atomic states couple and form molecular orbitals such as a low-energy orbital, called a bonding orbital, or a high-energy orbital, called an anti-bonding orbital. This is covalent bonding, which is distinctive of elements in group IVA of the periodic table, namely C, Si, Ge and Sn.

In terms of energy, both groups of orbitals are made of allowed bands for electrons. For a given interatomic distance, the population of allowed bands depends on the distance of orbitals s-p in the separated atom, and bonding-antibonding in chemical bonds in the crystal. The atom density is 10^{22} cm^{-3}.

Conduction phenomenon is the result of movement of electrons inside bands of bonding orbitals. Thus, an insulator only has completely filled bands or completely empty ones. A semiconductor has bands that allow for limited motion of electrons (these bands have a filling ratio of less than 25% or higher than 75%).

Allowed bands are classified as valence bands (which are full) and conduction bands (which are empty). There is a forbidden band between these bands, called a gap, which is associated with energy Eg from the thermal agitation of electrons. Eg is the energy needed to move an electron from a valence band to a conduction band. For semiconductors, Eg = 1.12 eV.

Single crystals of semiconductors (Si) are doped to increase or decrease electron density in allowed bands. Doping is ion implantation (I^2) of different atoms in the crystalline network. This may be:

– N-type, if implanted atoms play the role of donors, like in the case of P or As. An electron is liberated by thermal agitation (meV energy needed) and occupies a conduction band orbital;

– P-type, if implanted atoms play the role of acceptors, like in the case of B, Al or Ga. An electron moves from the valence band to the implanted atom through thermal agitation. Thus, a hole is created.

3.7.2. *Scharfetter–Gummel algorithm*

Hermann Gummel and Donald Lee Scharfetter used to work for Bell laboratories when they created their algorithm in 1950. This algorithm is a means to solve Poisson's current, transport and continuity equations in the P–N junction. The algorithm comes into play in the third phase of creating the program, that is, the equation-solving phases. Below, we will present the different steps that lead to the final equation system.

3.7.2.1. *Formulating equations*

We interpret conduction phenomena in a semiconductor using theories of quantum mechanics. By approximation, we deduce a transport equation, called the Boltzmann equation, for each type of particle.

3.7.2.2. *Normalizing equations*

Choosing variables that allow for discretization of the equation system is vital because it partly influences convergence properties of numerical methods.

In our case, we will use pseudo-Fermi levels to start the problem and carrier densities for solving it. By combining previously obtained equations, we get an equation system of the form [M] * [X] = [S], X being the unknown function, which we will have to solve in a space within which we will generate an appropriate mesh beforehand. The mesh used for discretizing equations must be sufficiently refined close to the junction in order to have convergence of the electron and hole current values. We will carry out this research in 1D. However, this will not affect the precision of the results.

3.7.2.3. *Solving equations, Scharfetter and Gummel algorithm*

The Scharfetter–Gummel method is the finite differences method. The current system is a 3.m.n coupled equation system, where m and n stand for the number of X and Y mesh points, of which potential, and electron and hole concentrations in each of the mesh points are unknown. This solution is quite delicate, as a large number of unknowns are introduced. In fact, it is necessary to have a dense mesh. The Gummel method of solving enables us to numerically solve equations of unidimensional geometry. This method has advantages such as a good numerical stability combined with a linear convergence and a low machine memory requirement; this is why we prefer this method to others.

3.7.2.4. *Discretization of equations*

We will numerically model equations that define potential (ψ) and pseudo-potentials (ψ_n and ψ_p) in the semiconductor.

We will use the following equation system:

1) $div(\mathrm{M_n} \cdot \exp(\varphi) \cdot \overrightarrow{grad}(\psi_n)) = \mathrm{GR}$

2) $div(M_p \cdot exp(-\varphi) \cdot \overrightarrow{grad}(\psi_p)) = -GR$

3) $div(\overrightarrow{grad}(\varphi)) = \exp(\varphi) \cdot \psi_n - \exp(-\varphi) \cdot \psi_p - DOP$

The equation is written in two dimensions:

$$div(M_n \cdot exp(\varphi) \cdot \overrightarrow{grad}(\psi_n)) = GR$$

$$\Leftrightarrow M_n \cdot div(\exp(\varphi) \cdot \overrightarrow{grad}(\psi_n)) = GR$$

$$\Leftrightarrow M_n \cdot \left[\exp(\varphi) \cdot div(\overrightarrow{grad}(\psi_n)) + \overrightarrow{grad}(\psi_n) \bullet \overrightarrow{grad}(\exp(\varphi))\right] = GR$$

$$\Leftrightarrow M_n \cdot \left[\exp(\varphi) \cdot \Delta(\psi_n) + \overrightarrow{grad}(\psi_n) \bullet \overrightarrow{grad}(\exp(\varphi))\right] = GR$$

$$\Rightarrow M_n \cdot \left[\exp(\varphi) \cdot \left(\frac{\partial^2 \psi_n}{\partial x^2} + \frac{\partial^2 \psi_n}{\partial y^2}\right) + \left(\frac{\partial \psi_n}{\partial x}\vec{i} + \frac{\partial \psi_n}{\partial y}\vec{j}\right) \bullet \left(\frac{\partial(\exp(\varphi))}{\partial x}\vec{i} + \frac{\partial(\exp(\varphi))}{\partial y}\vec{j}\right)\right] = GR$$

$$\Rightarrow M_n \cdot \left[\exp(\varphi) \cdot \left(\frac{\partial^2 \psi_n}{\partial x^2} + \frac{\partial^2 \psi_n}{\partial y^2}\right) + \exp(\varphi) \cdot \frac{\partial \varphi}{\partial x} \cdot \frac{\partial \psi_n}{\partial x} + \exp(\varphi) \cdot \frac{\partial \varphi}{\partial y} \cdot \frac{\partial \psi_n}{\partial y}\right] = GR$$

$$\Rightarrow M_n \cdot \exp(\varphi) \cdot \left[\frac{\partial^2 \psi_n}{\partial x^2} + \frac{\partial \varphi}{\partial x} \cdot \frac{\partial \psi_n}{\partial x} + \frac{\partial^2 \psi_n}{\partial y^2} + \frac{\partial \varphi}{\partial y} \cdot \frac{\partial \psi_n}{\partial y}\right] = GR \qquad \text{(I)}$$

But $\psi_n(x+h, y) = \psi_n(x,y) + h\frac{\partial \psi_n}{\partial x}(x,y) + \frac{h^2}{2}\frac{\partial^2 \psi_n}{\partial x^2}(x,y) + \circ(h^3)$

and $\psi_n(x-h, y) = \psi_n(x,y) - h\frac{\partial \psi_n}{\partial x}(x,y) + \frac{h^2}{2}\frac{\partial^2 \psi_n}{\partial x^2}(x,y) + \circ(h^3)$

Hence, $h^2\frac{\partial^2 \psi_n}{\partial x^2}(x,y) = \psi_n(x+h, y) + \psi_n(x-h, y) - 2\psi_n(x, y) + \circ(h^3)$ (II)

And also: $h\frac{\partial \psi_n}{\partial x}(x,y) = \psi_n(x, y) - \psi_n(x-h, y) + \circ(h^3)$ (III)

Applying the same method for the variable, we have:

$$k^2\frac{\partial^2 \psi_n}{\partial y^2}(x,y) = \psi_n(x, y+k) + \psi_n(x, y-k) - 2\psi_n(x, y) + \circ(k^3) \qquad \text{(IV)}$$

$$k\frac{\partial \psi_n}{\partial y}(x,y) = \psi_n(x, y) - \psi_n(x, y-k) + \circ(k^3) \qquad \text{(V)}$$

We substitute expressions (II), (III), (IV) and (V) into (I), which becomes:

$$\frac{1}{h^2}\left(\psi_n(x+h,y)+\psi_n(x-h,y)-2\psi_n(x,y)\right)+\frac{1}{h}\cdot\frac{\partial\varphi}{\partial x}\cdot\left(\psi_n(x,y)-\psi_n(x-h,y)\right)+$$
$$\frac{1}{k^2}\left(\psi_n(x,y+k)+\psi_n(x,y-k)-2\psi_n(x,y)\right)+\frac{1}{k}\cdot\frac{\partial\varphi}{\partial y}\cdot\left(\psi_n(x,y)-\psi_n(x,y-k)\right)=\frac{GR}{M_n\cdot\exp(\varphi)}$$

Three matrices can be associated with this equation: $\mathbf{A}\psi = \mathbf{F}$, with:

$$\mathbf{A}=\begin{pmatrix}
\frac{1}{h}\cdot\frac{\partial\varphi}{\partial x}-\frac{2}{h^2} & \frac{1}{h^2} & 0 & \cdots & \cdots & 0 & \frac{1}{k}\cdot\frac{\partial\varphi}{\partial y}-\frac{2}{k^2} & \frac{1}{k^2} & 0 & \cdots & \cdots & 0\\
\frac{1}{h^2}-\frac{1}{h}\cdot\frac{\partial\varphi}{\partial x} & \ddots & \ddots & \ddots & & \vdots & \frac{1}{k^2}-\frac{1}{k}\cdot\frac{\partial\varphi}{\partial y} & \ddots & \ddots & \ddots & & \vdots\\
0 & \ddots & \ddots & \ddots & \ddots & \vdots & 0 & \ddots & \ddots & \ddots & \ddots & \vdots\\
\vdots & \ddots & \ddots & \ddots & \ddots & 0 & \vdots & \ddots & \ddots & \ddots & \ddots & 0\\
\vdots & & \ddots & \ddots & \ddots & \frac{1}{h^2} & \vdots & & \ddots & \ddots & \ddots & \frac{1}{k^2}\\
0 & \cdots & \cdots & 0 & \frac{1}{h^2}-\frac{1}{h}\cdot\frac{\partial\varphi}{\partial x} & \frac{1}{h}\cdot\frac{\partial\varphi}{\partial x}-\frac{2}{h^2} & 0 & \cdots & \cdots & 0 & \frac{1}{k^2}-\frac{1}{k}\cdot\frac{\partial\varphi}{\partial y} & \frac{1}{k}\cdot\frac{\partial\varphi}{\partial y}-\frac{2}{k^2}
\end{pmatrix}$$

$$\boldsymbol{\psi}_n=\begin{pmatrix}\psi_x^1\\ \vdots\\ \psi_x^N\\ \psi_y^1\\ \vdots\\ \psi_y^N\end{pmatrix} \text{ and } \mathbf{F}=\begin{pmatrix}F_1\\ \vdots\\ \vdots\\ \vdots\\ \vdots\\ F_{2N}\end{pmatrix} \text{ with } F_i=\frac{GR}{M_n\cdot\exp(\varphi)},\ \forall i$$

By solving the system, we get the $\psi_n(x, y)$ expressions based on $\exp(\varphi)$.

We perform the same calculation for $\psi_p(x, y)$ based on equation: $div(M_p\cdot\exp(-\varphi)\cdot\overrightarrow{grad}(\psi_p)) = -GR$.

Thus, for each (x, y) point, we introduce the ψ_n and ψ_p values in the right-hand side of equation:

$$div(grad(\phi)) = \exp(\phi)\cdot\psi n - \exp(-\phi)\cdot\psi p - DOP$$

$$div(\overrightarrow{grad}(\varphi)) = \exp(\varphi)\cdot\psi_n - \exp(-\varphi)\cdot\psi_p - DOP \quad [3.173]$$

In this way, we have $div(\overrightarrow{grad}(\varphi))$ based on partial derivatives of φ, and by applying the finite differences method, we obtain the φ value at (x, y) points.

Solving **Poisson** and **div j = 0** (i.e. static) equations # This program was made using **MAPLE** with 2nd year Preparatory Level students of EURINSA, the European-French group of INSA, Lyon. This program uses the finite differences method. # **||*||** INITIALIZATION: PN JUNCTION (Diode) or MOS CAPCITOR **||*||**

```
Constant mesh:= false:
> # Solving Poisson and div j = 0 (i.e. static) equations
?
>
> # **||*||** INITIALIZATION: PN JUNCTION (Diode)
**||*||**
>
> Constant mesh:= false:
>
> # SI (MKSA) Units
```

3.8. Constants and data

Geometric data

```
> XSi:= 200.*10^(-9) : # height of junction
> Rp:= 0 : # position of doping peak with reference to
interface :Rp:= 0
> DeltaRp:= 0.5*10^(-8) : # standard deviation for
Gaussian doping
> PicDopage:= -10^26 : # Maximum doping in silicon
> DopageBulk:= 10^23 : # Substrate doping
```

Mesh and polarization date

```
> T:= 300 : # Temperature
```

```
> Vinitial:= 0 :# Initial voltage applied on surface
> Vfinal:= -0.25 : # Final voltage applied on surface
> NV:= 2 : # Number of polarizations
> Vb:= 0 : # Voltage applied on substrate 0V
> NNSi:= 201 : # Number of mesh points
> Raison:= 1.01 : # Ratio of geometric progression in x
meshes of silicon
```

Physical constants

```
> EpsVide:= 8.853*10^(-12): #Vacuum permittivity
> EpsSi:= 11.7*EpsVide: #Silicon permittivity
> ni:= 1.1*10^16: # Intrinsic carrier concentration (T =
300K)
> qElec:= -1.602*10^(-19): #Charge on an electron
> Mobn:= 1400*10(-4): #Electron mobility
> Mobp:= 900*10^(-4): #Hole mobility
> Boltzmann:= 1.381*10^(-23): # Boltzmann Constant
```

Numerical inspection data

```
> Arret:= 10^(-8): #Stopping test for relative value
Newton method - for Scharfetter Gummel method, root of
stopping test
> Maxit:= 20: #Max number of iterations, with or without
convergence
```

Auxiliary values

```
> UT:= abs(Boltzmann * T / qElec): #Thermal voltage
> Boltzmann:
> Diffusin = UT*Mobn:
>
> Diffusip = UT*Mobp:
> q:=qElec:
> MaxScharfetter:=Maxit:
> if Vinitial=Vfinal then DeltaV:= 0:
> else DeltaV:= (Vfinal-Vinitial)/(NV-1):
> end: #Step calculation
>
```

Doping definition: constant + Gaussian

```
> Dopage:= x -> DopageBulk - PicDopage / exp((x-
Rp)^2/DeltaRp^2 );
> plot(log10(abs(Dopage(x))),x=0..XSi,y=22..26);
>
```

```
>
```

Mesh

```
> NN:=NNSi:
> Dx:=array(1..NNSi,[]): #Dx is the table of delta x of
the uniform or non-uniform mesh.
> if (MaillageConstant=true)
> then hSi:=XSi/(NNSi-1): #Mesh size in silicon.
> for j from 1 to NNSi do Dx[j]:=hSi: od:
> else
> Sol:=solve({DopageBulk=-PicDopage / exp((x-
Rp)^2/DeltaRp^2 ),x>0},x):XJonction:= subs(Sol,x);
#Finding point where PN junction works
> NNMilieu:= ceil(NNSi/2):###if (modp(NNSi,2)=0)
> print ("Xjonction=",XJonction, "  ",
"NNMILIEU=",NNMilieu);
> ###ceil(NNSi/2)= NNSi/2: partie entiere +.5
> ###then NNMilieu:= NNSi/2: else NNMilieu:= NNSi/2 +.5:
end:
> DxIntG:= XJonction/sum('Raison^(i-
1)','i'=1..NNMilieu):
> DxIntD:= (XSi-XJonction)/sum('Raison^(i-
1)','i'=1..NNMilieu):
> for j from NNMilieu by -1 to 1 do
Dx[j]:=DxIntG*Raison^(NNMilieu-j+1): od:

> #for j from NNMilieu by -1 to 1 do
Dx[j]:=DxIntG*Raison^(NNMilieu-j-1): od:
> for j from NNMilieu+1 to NNSi do
Dx[j]:=DxIntD*Raison^(j-NNMilieu+1): od:
> for i from 1 by 50 to NN do
> print ("i =",i,"       Dx[i]",Dx[i]);
> od:
>
>
>
> ###DxInt:=XSi/sum('Raison^(i-1)','i'=1..NNSi);
> ###for j from 1 to NNSi do Dx[j]:=DxInt*Raison^(j-1):
od:
> end:
>
>
> NNMilieu:
> #XSi-XJonction:
> DxIntG:
```

```
> DxIntD:
> #print(Dx);
> X:=array(1..NNSi,[]): #X is the table of positions of
nodes in x of the mesh
> X[1]:=0:
> for j from 2 to NN-1 do X[j]:=X[j-1]+Dx[j-1]: od:
> X[NN]:=XSi:
> Dx[NN-1]:=XSi-X[NN-1]:
>
> listplot(X,connect=false,symbol=point);
>
>
>
> CoordX:= t -> X[t]*10^9:
> CoordY:= t -> 1:
>
>
plot([CoordX(t),CoordY(t),t=1..NN],0..XSi*10^9,0..2,styl
e=point);
>
plot([CoordX(t),CoordY(t),t=1..NNMilieu],0..XJonction*10
^9,0..2,style=point);
>
>
>
>
> #print(X):
>
>
```

Various initializations

Correction to electrostatic potential

```
> DPot:=Vector(NN,0):
> #DPot:=array(1..NN,[]):
```

Fermi level and carrier concentrations

```
> Matrice:=Matrix(NN,NN,0):
> Droite:=Vector(NN):
> MatriceDD:=Matrix(NN,NN,0):
> DroiteDD:=Vector(NN):
>
> Pot:=Vector(NN,0):
>
> Diag:=array(1..NNSi): SubDiag:=array(1..NNSi):
SuperDiag:=array(1..NNSi):
> DS_Eps_dx2iU:= array(1..NNSi-2):
```

```
> DS_Eps_dx2iL:= array(1..NNSi-2):
> #Trou:= Vector(NN,0):
> #Elec:=Vector(NN,0):
> Phip:=array(1..NN,[]):
> Elec:=array(1..NN,[]):
> Trou:=array(1..NN,[]):
> Phin:=array(1..NN,[]):
```

Save carrier concentrations for convergence test

```
> OldElec:=array(1..NN,[]): OldTrou:=array(1..NN,[]):
```

Doping mesh nodes

```
> DopageNet:=array(1..NN,[]):
> for i from 1 to NN do DopageNet[i]:= Dopage(X[i]):od:
for i from 1 by 20 to NN do
print ("i =",i,"        Dx[i]",Dx[i]);
od:

>
```

Initialization of all matrices

```
> for i from 1 to NN do
> Phin[i]:=0:
> Phip[i]:=0:

> if (i<NNMilieu) then
> Elec[i]:=DopageNet[i]:
> Trou[i]:=DopageBulk:
> else
> Elec[i]:=DopageNet[NNMilieu-1]-DopageBulk:
> Pot[i]:=0:

> Trou[i]:=DopageBulk:
> end if;
> OldElec[i]:=0:
> OldTrou[i]:=0:
> DPot[i]:=0:
> DroiteDD[i]:=0:
> Diag[i]:=0:
> SubDiag[i]:=0:
> SuperDiag[i]:=0:
> od:
>
> for i from 1 by 20 to NN do
> print ("i =",i," Elec[i]=",Elec[i],"
Trou[i]=",Trou[i]);
> od:
>
```

```
> #Initial electrostatic potential
> if DopageNet[NN]>0 then
> PsiB:=Vb-UT*ln(abs(DopageNet[NN])/ni):
> else
> PsiB:=Vb+UT*ln(abs(DopageNet[NN])/ni):
> end if:
> PsiB;
>
>
>
> for i from 1 to NN do Pot[i]:=PsiB: od:

> #Surface polarization
> V:=array(1..NV): for i from 1 to NV do V[i]:=
Vinitial+(i-1)*(Vfinal-Vinitial)/(NV-1): od:
> print(NV,"tension(s) applique(s) : ",V);
>
> print ("Initialisation des matrices");
>
>
```

Laplace

```
> Laplacien:= proc() local i;global DS_Eps_dx2iU,
DS_Eps_dx2iL, Diag, SuperDiag, SubDiag, Dx, NN, NNSi,
EpsSi;
> print("Bonjour: ici mon laplacien");
>
> for i from 1 to NNSi-2 do
> DS_Eps_dx2iU[i]:= (EpsSi*2/(Dx[i+1]*(Dx[i]+Dx[i+1]))):
> DS_Eps_dx2iL[i]:= (EpsSi*2/(Dx[i]*(Dx[i]+Dx[i+1]))):
> od:
> Diag[1]:=1:
> SubDiag[1]:=0:
> SuperDiag[1]:=0:
>
> for i from 2 to NN-1 do
> Diag[i]:= -DS_Eps_dx2iU[i-1] -DS_Eps_dx2iL[i-1]:
> SubDiag[i]:=DS_Eps_dx2iL[i-1]:
> SuperDiag[i]:=DS_Eps_dx2iU[i-1]:
> #SuperDiag[i]:=DS_Eps_dx2iU[i-1]:

> od:
>
> Diag[NN]:=1:
> SubDiag[NN]:= 0:
```

```
> SuperDiag[NN]:=0:
> for i from 1 by 50 to NN do
> print ("Diag    " , Diag[i]);
> print ("SubDiag    " , SubDiag[i]);
> print ("SuperDiag    " , SuperDiag[i]);
> od:
> end proc:
>
>
>
```

Right hand side ()

```
> MbDroite:= proc () global Droite, PsiS,PsiB,Pot,NN, q,
ni,DS_Eps_dx2iL, DS_Eps_dx2iU,DopageNet, Phin, Phip, UT;
local i,k;
```

Ohmic contracts: PsiS, PsiB(fixe),

```
> Droite[1]:= PsiS-Pot[1]:print("Droite[1]:= PsiS-
Pot[1]=",Droite[1]);
> print("Potentiel du contact ohmique en
surface=",Droite[1]);
> Droite[NN]:=PsiB-Pot[NN]:print("Droite[NN]:= PsiS-
Pot[NN]=",Droite[NN]);
> print("Potentiel du contact ohmique du
substrat=",Droite[NN]);
```

Calculating second part in silicon for Newton part

```
> for i from 2 to NN-1 do
> Droite[i]:= (DS_Eps_dx2iL[i-1])*(Pot[i]-Pot[i-1]) +
(DS_Eps_dx2iU[i-1])*(Pot[i]-Pot[i+1]) - q*( ni*( exp(
(Pot[i]-Phin[i])/UT )- exp( (Phip[i]-Pot[i])/UT ) ) +
DopageNet[i] ):
> od:
> for k from 1 by 1 to 10 do
> print("k=",k,"Droite:",Droite[k]);
> od:
> print ( "fin du calcul de Droite ");
> end proc:
```

Poisson

```
> Poisson:= proc() global Arret, Maxit,Droite,
Diag,Matrice,DPot,Pot,Phin,Phip,SuperDiag,SubDiag,
NN,NNSi, UT,q,ni, DS_Eps_dx2iU, DS_Eps_dx2iL, Dx, EpsSi,
PsiS,PsiB,DopageNet ; local k,i, Erreur, IterNewton;
> print("Lancement de Poisson");
```

Boundary conditions: Ohmic contacts: PsiS, PsiB(fixe),
Linearization --> Newton-Raphson
Constructing and solving linear system
Convergence test

```
>

> Erreur:=100:
> IterNewton:=0:
> # print("il faut (Erreur>Arret) and
(IterNewton<Maxit)");
> # printf("Erreur:%a - Arret:%a - IterNewton:%a -
#Maxit:%a",Erreur,Arret,IterNewton,Maxit);
> #if( (Erreur>2)and (IterNewton<100000000000)) then
> while( (Erreur>Arret) and (IterNewton<Maxit) ) do
> print("Entree dans la boucle de Poisson");
```

Calculating 3 diagonals of the linear system for the Laplace part

```
> print("Lancement prochain de Laplacien");
> Laplacien():
```

Correction to diagonal of the linear system for Newton part

```
> print("Laplacien effectue");
> print("Lancement Boucle: Mat, Diag, Droite");
> for k from 2 to NN-1 do
> Diag[k]:=Diag[k]+( (q*ni)/UT)* ( exp( (Pot[k]-
Phin[k])/UT ) + exp( (Phip[k]-Pot[k])/UT ) ):
> od:
> print("Boucle de stockage de Diag final dans Mat");
> for k from 1 to NN-1 do
>
>
```

Full matrix storage

```
> Matrice[k,k]:=Diag[k]:
> Matrice[k,k+1]:=SuperDiag[k]:
> Matrice[k+1,k]:=SubDiag[k+1]:
> od:
> Matrice[NN,NN]:=Diag[NN]:
>
> print("Matrice stockee, enfin normalement");
>
```

Calculating second part of Simeon-Denis

```
> print("Calcul du second membre de Fish");
> MbDroite():
```

Solving linear system

```
> #print("!!!!!!!
=",Droite[1],Droite[2],Droite[3],Droite[101],Droite[201]
);
```

```
> with(LinearAlgebra,LinearSolve);
> #print(Matrice);
> print("Matrix sub,diag et super, puis  Droite");
> for i from 2 by 1 to 10 do
> printf("%g | %g | %g || %g", Matrice[i,i-
1],Matrice[i,i],Matrice[i,i+1],Droite[i]);print("");od:
> DPot:= LinearSolve(Matrice,Droite):
> print("inversion achevee chez Poisson");
> #print(DPot);
>
> #print(Matrice);
> #print(Droite);
```

Updating potential, calculating error in Poisson

```
> for k from 1 to NN do
> Pot[k]:= Pot[k]+DPot[k]:
> od:
> for k from 1 by 50 to NN do
> print("k=",k,"    ","DPot=",DPot[k],"
","Pot=",Pot[k]);
> od:

> Erreur:= sqrt( sum('DPot[j]^2','j'=1..NN)/NN ):
> print ("Erreur moyenne sur DeltaPotentiel:", "    ",
Erreur);
> IterNewton:=IterNewton+1:
> print("          Pas iteration dans Poisson:IterNewton=
",IterNewton);
>
```

End of iteration

```
> od:

> print("Convergence Poisson en ",
IterNewton,"iterations");
> print ("  FIN    DE     POISSON          ");
>
> end proc:
>
>

>
>
>
```

Transport

```
> Transport:= proc() global porteur,Pot,MatriceDD,
DroiteDD,
Diag,SubDiag,SuperDiag,DopageNet,Elec,Trou,Phin,Phip,Mob
n,Mobp,UT, NNSi, ni: local TrouB,TrouS,ElecB,ElecS,i:
> print( " TRANSPORT - electrons  et  trous : demarrage
, porteur=",porteur);
>
```

Calculating matrix terms in the form of 3 diagonals and MatriceDD report

```
> if (porteur = 1) then
  DivBJp():
else
  DivBJn():
end if:
> for i from 1 while i <= NNSi do
  MatriceDD[i,i]:= Diag[i]:
end do:
> for i from 1 while i <= NNSi-1 do
  MatriceDD[i+1,i]:= SubDiag[i+1]:
  MatriceDD[i,i+1]:= SuperDiag[i]:
end do:
```

Ohmic contacts: surface and substrate

```
> if (DopageNet[1] > 0) then
  ElecS:= DopageNet[1]:
 TrouS:= ni^2/ElecS:
else
  ElecS:= -DopageNet[1]:
  TrouS:= ni^2/ElecS:
end if:
> if (DopageNet[NN] > 0) then
  TrouB:= DopageNet[NN]:
  ElecB:= ni^2/TrouB:
else
  ElecB:= -DopageNet[NN]:
 TrouB:= ni^2/ElecB:
end if:
```

Solving linear system for transport

```
> #print("Matrice finale", Matrice):
> #print("Droite finale", DroiteDD):

> if (porteur = -1) then
  print("On resout le systeme lin.pour les electrons:
porteur=", porteur,"  Elec=", ElecS, "ElecB=",ElecB);
  DroiteDD[1]:= ElecS:
  DroiteDD[NNSi]:= ElecB:
```

```
  for i from 2 while i < NNSi do
   DroiteDD[i]:= 0:
  end do:
  Elec:= LinearSolve(MatriceDD, DroiteDD):
  for i from 1 to NN do
   #Elec[i]:=Elec[i]:
   Phin[i]:=Pot[i]-UT*ln(Elec[i]/ni):
  end do:

for i from 1 by 25 to NN do
   print(" Elec[i]" , Elec[i]);
   print(" Phin[i] ", Phin[i]);
  end do:
 else
  print("On resout le systeme lin. pour les trous,
porteur=", porteur, " TrouS=", TrouS, "  TrouB=",TrouB):
  DroiteDD[1]:= TrouS:
  DroiteDD[NNSi]:= TrouB:
  for i from 2 while i < NNSi do
   DroiteDD[i]:= 0:
  end do:
  Trou:= LinearSolve(MatriceDD, DroiteDD):
  for i from 1 to NN do
  #  Trou[i]:=Trou[i]:
   Phip[i]:=Pot[i]+UT*ln(Trou[i]/ni):
  end do:
 for i from 1 by 25 to NN do
   print(" Trou[i]" , Trou[i]);
   print(" Phip[i] ", Phip[i]);
  end do:

end if:
>
> end proc:
>
>
>
```

Bernoulli (u)

```
> #Bernou:= proc(u) global UT;
> #print ("  DEBUT   DE      BERNOULLI     ","u=", u);

> #if (abs(u/UT) < 10^(-10)) then
> #(1-u/2/UT):
```

```
> #else
> #(u/UT)/(exp(u/UT)-1):
> #print ("   FIN    DE      BERNOULLI     ","Bernou",
Bernou(u));
> #end:end proc:
```

Div(Jn)

```
> DivBJn:= proc() global UT,NN, Mobn, Pot,Diag,
SuperDiag, SubDiag:local i, u, v,u1,v1:
> Diag[1]:=1:
> Diag[NN]:=1:
> for i from 2 to NN-1 do
> u:=Pot[i]-Pot[i+1]:
> if (abs(u/UT) < 10^(-10)) then
> u1:=(1-u/2/UT):
> else
> u1:=(u/UT)/(exp(u/UT)-1):
> end if;

> v:=Pot[i]-Pot[i-1]:

> if (abs(v/UT) < 10^(-10)) then
> v1:=(1-v/2/UT):
> else
> v1:=(v/UT)/(exp(v/UT)-1):
> end if;

> Diag[i]:=(u1+v1)*Mobn:
> end do:
> SubDiag[1]:=0:
> SubDiag[NN]:=0:
> SuperDiag[1]:=0:
> SuperDiag[NN]:=0:

> for i from 2 to NN-1 do
> u:=Pot[i-1]-Pot[i]:
> if (abs(u/UT) < 10^(-10)) then
> u1:=(1-u/2/UT):
> else
> u1:=(u/UT)/(exp(u/UT)-1):
> end if;

> v:=Pot[i+1]-Pot[i]:
```

```
> if (abs(v/UT) < 10^(-10)) then
> v1:=(1-v/2/UT):
> else
> v1:=(v/UT)/(exp(v/UT)-1):
> end if;

> SubDiag[i]:=-u1*Mobn:
> SuperDiag[i]:=-v1*Mobn:
> end do:

> print(" fin  de  Div(Jn)");
> end proc:
>
```

Div(Jp)

```
> DivBJp:= proc() global UT,NN, Mobp, Pot, Diag,
SuperDiag, SubDiag:local i, u, v, u1,v1;
> Diag[1]:=1:
> Diag[NN]:=1:
> for i from 2 to NN-1 do
> u:=Pot[i+1]-Pot[i]:
> if (abs(u/UT) < 10^(-10)) then
> u1:=(1-u/2/UT):
> else
> u1:=(u/UT)/(exp(u/UT)-1):
> end if;

> v:=Pot[i-1]-Pot[i]:
> if (abs(v/UT) < 10^(-10)) then
> v1:=(1-v/2/UT):
> else
> v1:=(v/UT)/(exp(v/UT)-1):
> end if;

> Diag[i]:=(u1+v1)*Mobp:
> od:

> SubDiag[1]:=0:
> SubDiag[NN]:=0:
> SuperDiag[1]:=0:
> SuperDiag[NN]:=0:

> for i from 2 to NN-1 do
> u:=Pot[i]-Pot[i-1]:
```

```
> if (abs(u/UT) < 10^(-10)) then
> u1:=(1-u/2/UT):
> else
> u1:=(u/UT)/(exp(u/UT)-1):
> end if;

> v:=Pot[i]-Pot[i+1]:
> if (abs(v/UT) < 10^(-10)) then
> v1:=(1-v/2/UT):
> else
> v1:=(v/UT)/(exp(v/UT)-1):
> end if;

> SubDiag[i]:=-u1*Mobp:
> SuperDiag[i]:=-v1*Mobp:
> od:
> print("    fin  de  Div(Jp)");
> end proc :
>
```

PlotPot

```
> PlotPot:=proc() local
i,potarray,phinarray,phiparray,P1,P2,P3; global
NN,X,Pot,Phin,Phip;

>
> potarray:= array(1..2,1..NN):
> phinarray:= array(1..2,1..NN):
> phiparray:= array(1..2,1..NN):
> for i from 1 to NN do potarray[1,i]:= X[i] end do:
> for i from 1 to NN do potarray[2,i]:= Pot[i] end do:
> for i from 1 to NN do phinarray[1,i]:= X[i] end do:
> for i from 1 to NN do phinarray[2,i]:= Phin[i] end do:
> for i from 1 to NN do phiparray[1,i]:= X[i] end do:
> for i from 1 to NN do phiparray[2,i]:= Phip[i] end do:
> with(plots):
> P1:=listplot(potarray):
> P2:=pointplot(phinarray, color=blue,symbol=point):
> P3:=pointplot(phiparray, color=red,symbol=point):
> display(P1,P2,P3,title="Potentiel Electrique (V)");
> end proc():
```

SomCharge

```
> # SomCharge:=proc() local i; global # #
NN,NNOx,ni,Pot,Phin,UT,Phip,Elec,Trou,Charge,Chargen,Cha
rgep,Dx;
> # for i from 1 to NNOx do Elec[i]:= 0 end do:
> # for i from NNOx+1 to NN do Elec[i]:= ni*exp((Pot[i]-
Phin[i])/UT) end do:
> # for i from 1 to NNOx do Trou[i]:= 0 end do:
> # for i from NNOx+1 to NN do Trou[i]:=
ni*exp((Phip[i]-Pot[i])/UT) end # do:
> # Chargen[jV]:=0;
> # Chargep[jV]:=0;
> # for i from NNOx+1 to NN-1 do #
Chargen[jV]:=Chargen[jV]+Elec[i]+Elec[i+1]*Dx[i]/2 end
do:
> # for i from NNOx+1 to NN-1 do #
Chargep[jV]:=Chargeo[jV]+Trou[i]+Trou[i+1]*Dx[i]/2 end
do:
> # Charge:=Chargen-Chargep:
> # end proc():
```

PlotChargeX

```
> # PlotChargeX:=proc() local #
i,n,doparray,elecarray,trouarray,P4,P5,P6; global #
NN,X,DopageNet,Elec,Trou;
>
> # doparray:= array(1..2,1..NN):
> # elecarray:= array(1..2,1..NN):
> # trouarray:= array(1..2,1..NN):
> # for i from 1 to n do doparray[1,i]:= X[i] end do:
> # for i from 1 to n do doparray[2,i]:=
log10(abs(DopageNet[i])) end do:
> # for i from 1 to n do elecarray[1,i]:= X[i] end do:
> # for i from 1 to n do elecarray[2,i]:=
log10(abs(Elec[i])) end do:
> # for i from 1 to n do trouarray[1,i]:= X[i] end do:
> # for i from 1 to n do trouarray[2,i]:=
log10(abs(Trou[i])) end do:
> # with(plots):
> # P4:=listplot(doparray):
> # P5:=pointplot(elecarray, color=blue,symbol=point):
> # P6:=pointplot(trouarray, color=red,symbol=point):
> # display(P4,P5,P6,title="Dopage, lectrons et trous
(Log)");
```

```
> # end proc():
```

PlotChargeV

```
> # PlotChargeV:=proc() local i,chearray,chparray,P7,P8;
global # V,jV,Chargen,Chargep;
> # chearray:= array(1..2,1..jV):
> # chparray:= array(1..2,1..jV):
> # for i from 1 to jV do chearray[1,i]:= V[i] end do:
> # for i from 1 to jV do chearray[2,i]:=
log10(Chargen[i]) end do:
> # for i from 1 to jV do chparray[1,i]:= V[i] end do:
> # for i from 1 to jV do chparray[2,i]:=
log10(Chargep[i]) end do:
> # with(plots):
> # P7:=listplot(chearray, color=red):
> # P8:=pointplot(chparray, color=blue,symbol=point):
> # display(P7,P8,title="Q(V)");
> # end proc():
>
>
> with(LinearAlgebra):
>
>
>
>
```

Beginning of main program

Loop over voltage

```
> print("Debut de PROGRAMME");
```

```
> for jV from 2 to NV do
```

Calculating surface potential PsiS

```
> if DopageNet[1]>0
> then PsiS:= V[jV]-UT*ln(DopageNet[1]/ni):
> else PsiS:= V[jV]+UT*ln(-DopageNet[1]/ni):
> fi:
```

Rectifying potential to take surface potential into account

```
> for i from 1 to NN do
> if DopageNet[1]*DopageNet[i]>0
> then Pot[i]:= (PsiS+PsiB)/2:
> #then Pot[i]:= (0)/2:
```

```
> end if;
> od:
> print("Vsurface : ",PsiS," VBulk : ",PsiB);
>
```

Beginning of Scharfetter-Gummel iteration for Poisson and Transport decoupling

```
> ErreurE:=10000:
> iterScharfetter:= 0:
> while ((ErreurE>sqrt(Arret)) and
(iterScharfetter<MaxScharfetter) ) do
>
> Poisson():
> iterScharfetter:= iterScharfetter + 1:
```

Solving the two transport equations

```
> print("      Boucle sur decouplage,iterScharfetter
",iterScharfetter);

> #convert(Elec,vector):
> #convert(Trou,vector):
> porteur:=-1:
> Transport():
> print("Equation de transport des electrons resolue");
> porteur:=1:

> Transport():
> print("Equations de transport des trous resolue");
```

Calculating error in Poisson decoupling and loop

```
> ErreurDeltaPorteurs:= sqrt(sum('(Elec[i]-
OldElec[i])^2', 'i'=1..NN)/NN)+sqrt(sum('(Trou[i]-
OldTrou[i])^2','i'=1..NN)/NN):
> Norme:= sqrt( sum('(Elec[i])^2','i'=1..NN)/NN ) +
sqrt( sum('(Trou[i])^2','i'=1..NN)/NN ):
> print("Erreur",ErreurDeltaPorteurs, "Norme=",Norme);
>
> ErreurE:= ErreurDeltaPorteurs/Norme:
> OldElec:= Elec: OldTrou:= Trou:
> for i from 1 by 50 to NN do
> #print("OldElec=",OldElec[i],"
","Elec[i]=",Elec[i]," verif: crassement de matrice");
> #print("OldTrou=",OldTrou[i],"
","Trou[i]=",Trou[i]," verif: crassement de matrice");
> od:
> print("Erreur moyenne sur porteurs=",ErreurE,
"IterScharfetter ",iterScharfetter);
> od:
```

End of Scharfetter-Gummel iteration

```
> PlotPot():

> # SomCharge:
```
Graphs
```
> # PlotChargeX:
> PlotPot():
> # PlotChargeV:
> od:
```

End of loop over voltage

End of Program
```
> print("Fin prgm");
>
```

We give also an example of solving an electronic problem (calculating n, p and V in a diode, then continuing on the bipolar transistor) using the **finite element method** thanks to the numerical calculation software MATLAB [x].

This program was created using **MATLAB** with second-year Preparatory-Level students of AMERINSA, the Latin-French Bachelor studies group of INSA, Lyon. This program uses the finite elements method.

The program has the following individual files:

-**Struct**, which defines the diode's geometry as well as the mesh used.

-**Initi**, which defines the different parts of the diode (emitter, base and collector) and creates doping matrices.

-**Dopage**, which dopes different parts of the diode.

-**CD-FRT**, which defines the system's boundary conditions.

-**Solu** and **Solution finale**, finalizing the problem, calculating:

---N: electron density

---P: hole density

---V: potential (so electrical field): to finalize.

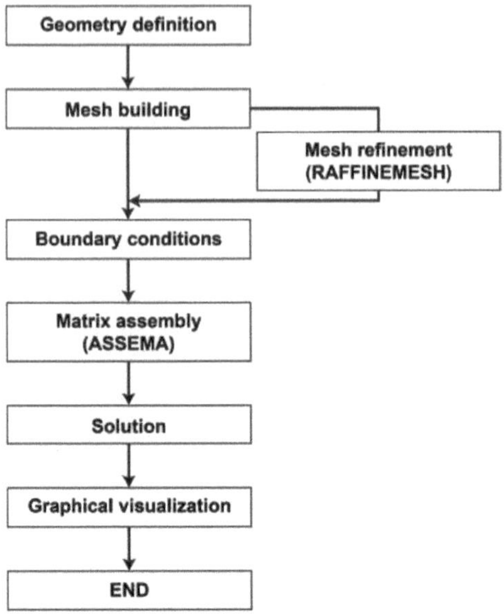

Flowchart of the finite element method under MATLAB

Figure 3.71. *General FEM flow chart*

Example of Poimesh:

. tic;

Beginning of program

Problem details:

ld=3.39296E-5;

Diode length

x1=0;x2=1;

y1=0;y2=1;

Two-point coordinates (0,0) and (1,1)

g1=[2 x1 x2 y1 y1 1 0];

g2=[2 x2 x2 y1 y2 1 0];

g3=[2 x2 x1 y2 y2 1 0];

g4=[2 x1 x1 y2 y1 1 0];

1 and 0 define the left and right regions and thus, the flow direction.

geo=[g1' g2' g3' g4'];

geo is the matrix that defines the geometry of the problem, which is a square in this case.

Mesh:

Two meshes are possible: the classic initmesh or the Poisson mesh that constructs triangles.

%[p,e,t]=initmesh(geo);

[p,e,t]=poimesh(geo,2,2);

Poimesh arguments are the geometry matrix, and number of columns and lines, respectively.

In addition to building a mesh for the Delaunay mathematical method, the initmesh function (“initialize mesh”) or poimesh function refer to three matrices: **p, e and t**.

Matrix **p** (point matrix) has x and y coordinates for all points (nodes) in the mesh, from left to right and from top to bottom (in the order of Figure 3.72). **E** is the edge matrix and contains 7 rows:

– Rows 1 and 2: indices of the points connected to each other (for indexing, see example: "maillage poisson.bmp").

– Rows 3 and 4: varying vector coordinates for the edge. These coordinates are often either x or y but never both at the same time (example: 0 0.5).

– Row 5: indicates the number of the edge segment on which the “varying vectors” are found (numbered in the sweep direction (there are 4).

– Rows 6 and 7: indicate the number of right- and left-hand side subdomains, row 6 containing the left-hand side subdomains and row 7 the right-hand side ones.

t matrix (triangle matrix):

For each column, the first 3 rows contain the 2 nodes defining a triangle, starting from the bottom-left point and going forward in the trigonometrical direction (however, the order of triangles seems random on the Poisson mesh).

pdemesh(p,e,t):

This function only plots the mesh specified by the three abovementioned matrices.

disp :(mesh end):

toc;

Example of a very simple mesh defined above:

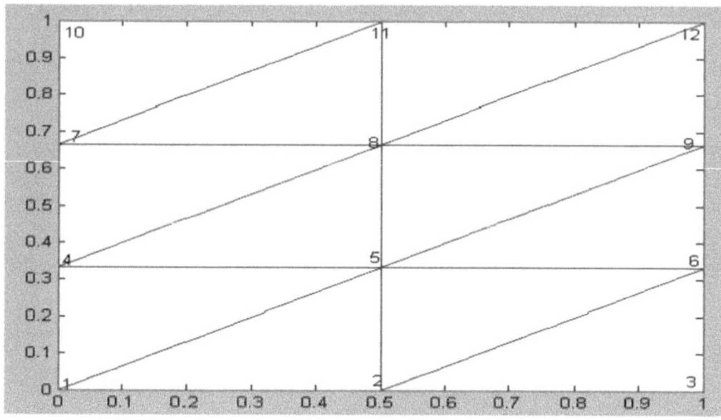

Figure 3.72. *A created simple mesh (MATLAB). For a color version of this figure, see www.iste.co.uk/gontrand/nanoelectronics.zip*

P matrix:

$$\begin{bmatrix} 0 & 0{,}5 & 1 & 0 & 0{,}5 & 1 & 0 & 0{,}5 & 1 & 0 & 0{,}5 & 1 \\ 0 & 0 & 0 & 0{,}33 & 0{,}33 & 0{,}33 & 0{,}66 & 0{,}66 & 0{,}66 & 1 & 1 & 1 \end{bmatrix}$$

e matrix:

$$\begin{bmatrix} 1 & 2 & 3 & 6 & 9 & 12 & 11 & 10 & 7 & 4 \\ 2 & 3 & 6 & 9 & 12 & 11 & 10 & 7 & 4 & 1 \\ 0 & 0{,}5 & 0 & 0{,}33 & 0{,}66 & 0{,}5 & 0{,}5 & 0 & 0{,}33 & 0{,}66 \\ 0{,}5 & 1 & 0{,}33 & 0{,}66 & 1 & 1 & 1 & 0{,}33 & 0{,}66 & 1 \\ 1 & 1 & 2 & 2 & 2 & 3 & 3 & 4 & 4 & 4 \\ 1 & 1 & 1 & 1 & 1 & 1 & 1 & 1 & 1 & 1 \\ 0 & 0 & 0 & 0 & 0 & 0 & 0 & 0 & 0 & 0 \end{bmatrix}$$

t matrix:

$$\begin{bmatrix} 7 & 8 & 1 & 4 & 7 & 5 & 2 & 5 & 2 & 8 & 4 & 1 \\ 11 & 9 & 5 & 8 & 8 & 6 & 3 & 9 & 6 & 12 & 5 & 2 \\ 10 & 12 & 4 & 7 & 11 & 9 & 6 & 8 & 5 & 11 & 8 & 5 \\ 1 & 1 & 1 & 1 & 1 & 1 & 1 & 1 & 1 & 1 & 1 & 1 \end{bmatrix}$$

A

Initialization

- Initializing calculations
- Defining problem

tic;

Problem details:

Identifying node xi and yi

x=p(1,:);

y=p(2,:);

matrice(numlign,nbcol) refers to the numlign row and nbcol columns of the concerned matrix. The notation ":" means a complete sweep.

After these two instructions, we have 2 matrices of 1 dimension each with x and y coordinates of all mesh nodes.

I=1:size(t,2);

Function returns the size of the object (for a matrix, this refers to the number of columns) of the triangle matrix. 2 is the convention that returns the number of columns of the matrix and 1 returns the number of rows.

xmoy=(x(t(1,:))+x(t(2,:))+x(t(3,:)))/3;

ymoy=(y(t(1,:))+y(t(2,:))+y(t(3,:)))/3;

Let us consider the term x(t(1,:)), x is an application (represented by x matrix defined at the beginning). x(index) returns the x coordinates of this node. t(1,:) is the row matrix corresponding to the first row of the t triangle matrix (that is, the index of the first node of each triangle). And thus, x(t(1,:)) returns a matrix corresponding to the abscissae of each node of the first row of the matrix.

Similarly, **x(t(2,:))** and **x(t(3,:))** are for the 2nd and 3rd rows of t matrix, that is, for the 2nd and 3rd nodes of each triangle.

Then, **xmoy** calculates the average abscissa of each triangle and xmoy is the row matrix for which the number of triangles is the number of columns.

Allocations:

nbtrig=size(t,2);

returns the column number of t (or the triangle number).

dop=zeros(1,nbtrig);

dop is a zero matrix whose dimension is one times nbtrig

dopb=zeros(1,nbtrig);

dope=zeros(1,nbtrig);

for m=1:nbtrig

(<=>for(int m=1;m<=nbtrig;m++)

carries out different doping following the current domain (domain of either base, emitter or collector).

if(t(4,m)==3 | t(4,m)==2 | t(4,m)==1)

t(4,m) (here m goes from 1 to nbtrig) and t(4,m) return the number of the domain where the concerned triangle is.

dopb(1,m)=-csb*exp(-d0b*(ymoy(1,m))-y4.^2);

dopb(1,m) that is, the value of the first row and m^{th} column will be replaced by -csb*exp(-d0b*(ymoy(1,m))-y4.^2).

```
dop(1,m)=dopb(1,m);
```

These new values are copied into the dop matrix.

```
end
if(t(4,m)==4 | t(4,m)==5)
dope(1,m)=cse*exp(-d0e*(ymoy(1,m)-y6).^2);
dop(1,m)=dope(1,m);
end
end
disp('fin de l''initialisation')
```

Doping at node points:

```
Nnb1p=find((y<y4+10*eps)&(y>y1-10*eps)&(x<x6+10*eps)&(x>x1-10*eps));
</x6+10*eps)&(x></y4+10*eps)&(y>
```

Selecting base nodes; returns position indices of these nodes.

```
Nnep=find((y<y6+10*eps)&(y>y4+10*eps)&(x<x4+10*eps)&(x>x3-10*eps));
</x4+10*eps)&(x></y6+10*eps)&(y>
```

Selecting emitter nodes; returns position indices of these nodes.

Initializing node doping matrices:

```
nbp=size(p,2);
dopb10=zeros(1,nbp);
dope0=zeros(1,nbp);
```

It builds a one-row matrix with nbp columns (where nbp: number of columns of p matrix = number of points) and fills this matrix of zero.

dop0=zeros(1,nbp);

Emitter:

dope0(Nnep)=cse*exp(-d0e*((y(Nnep)-y6)).^2);

dop0(Nnep)=dope0(Nnep);

Base:

dopb(Nnb1p)=-csb*exp(-d0b*(y(Nnb1p)-y4).^2);

dop0(Nnb1p)=dopb(Nnb1p);

Calculating properties of triangles:

dop01=dop0';

pdesurf(p,t,dop01) the function pdesurf three-dimensionally traces the surface based on the triangle and points matrices and the desired doping.

dop0moy=(dop0(t(1,:))+dop0(t(2,:))+dop0(t(3,:)))/3;

Doping

Calculating the doping profile in the base-emitter junction:

tic;

clc; initialization (reset inputs and outputs to zero)

clear; reset all variables to zero

clg; reset design space to zero.

Data:

ld=3.39296E-5;

diode width

y1=0E-12/ld;

y4=200E-12/ld;

y6=730E-12/ld;

y1, y4 and y6 are coordinates of points used to trace the Gaussian curve of doping.

ni=1.45E10;

Number of carries doped unintentionally.

csb=1E18/ni;

cse=1E19/ni;

cepi=8E16/ni;

Carrier concentrations in base and emitter.

d0e=-1/(y4-y6)^2*[log(csb/cse)];

d0b=-1/(y1-y4)^2*[log(cepi/csb)];

x1=0E-12/ld;

x2=500E-12/ld; (valeur non utilisée)

x3=1250E-12/ld;

x4=3150E-12/ld;

x5=3900E-12/ld; (valeur non utilisée)

x6=4400E-12/ld;

x1, x3, x4 and x6 are coordinates of points used to trace the Gaussian curve of doping.

Chart of E/B diode doping:

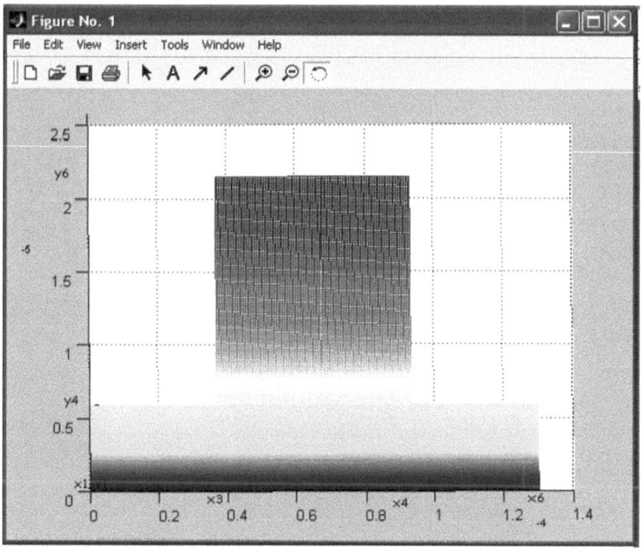

Figure 3.73. *Meshing of the MOS. For a color version of this figure, see www.iste.co.uk/gontrand/nanoelectronics.zip*

Emitter doping profile:

yi0=y6;

yf0=y4;

np=100;

Number of carriers

pa0=(yf0-yi0)/(np-1);

Carrier width (y axis):

Y1=[yi0:pa0:yf0];

```
ce=[1E18/ni:pa0:1E19/ni];
xi0=x3;
xf0=x4;
np=100;
h0=(xf0-xi0)/(np-1);
```

Carrier height (x axis)

```
xe=[xi0:h0:xf0];
for i=1:np
for j=1:np
ce(i,j)=cse*exp(-d0e*(Y1(i)-y6)^2);
end;
end;
```

Base doping profile:

```
yi1=y4; beginning of base
yf1=y1; end of base
np=100;
pa1=(yf1-yi1)/(np-1);
```

Carrier width

```
Y2=[yi1:pa1:yf1];
l0=100;
p0=(x6-x1)/(l0-1);
xb1=[x1:p0:x6];
for i=1:np
```

```
for j=1:l0
cb1(i)=csb*exp(-d0b*(Y2(i)-y4)^2);
```

Gaussian profile of carrier concentration in base:

```
cb1(i,j)=cb1(i);
j=j+1;
end;
end;

mesh(xe,Y1,log10(ce));
hold on
mesh(xb1,Y2,log10(cb1));
end
```

Boundary conditions

Dirichlet boundary conditions:

```
tic;
```

Conditions for N carriers:

```
Dir1N=2./(abs(dop0(Nn1))+sqrt(4+abs(dop0(Nn1)).^2));
Dir2N=(dop0(Nn2)+sqrt(4+(dop0(Nn2).^2)))/2;
```

Conditions for P carriers:

```
Dir1P=(abs(dop0(Nn1))+sqrt(4+(abs(dop0(Nn1)).^2)))/2;
Dir2P=2./(abs(dop0(Nn2))+sqrt(4+abs(dop0(Nn2)).^2));
```

Potential conditions:

```
ut=0.0258;
phib=0.0/ut;
phie=0.4/ut;
Dir1phi=phib-(log(Dir1P));
Dir2phi=phie+(log(Dir2N));
```

N-type pseudo-Fermi level conditions:

```
Dir1phino=phib;
Dir2phino=phie;
```

Pseudo-Fermi level conditions: final solution:

```
Dir1EXIn=exp(-phib);
Dir2EXIn=exp(-phie);
```

P-Type:

```
Dir1phipo=phib;
Dir2phipo=phie;
```

Pseudo-Fermi level conditions: final solution:

Dir1EXIp=exp(phib);

Dir2EXIp=exp(phie);

disp('frontiéres de dirichlet') Write what is between apostrophes (= System.out.println();)

Dirichlet boundary points:

Nn1=find((y<y1+10*eps)&(y>y1-10*eps)&(x<x6+10*eps)&(x>x1-10*eps));

Nn2=find((y<y6+10*eps)&(y>y6-10*eps)&(x<x4+10*eps)&(x>x3-

10*eps));</x4+10*eps)&(x></y6+10*eps)&(y></x6+10*eps)&(x></y1+10*eps)&

(y>

Selects indices of the boundary nodes and puts them in a matrix.

toc ;

Appendices

Two-dimensional Poisson equation, like:

$$-\frac{d^2u(x,y)}{dx^2}-\frac{d^2u(x,y)}{dy^2}=f(x,y)\ (2.1)\text{ on } D. \quad [3.174]$$

We apply Dirichlet and Neumann conditions to define the boundary equations:

u(x,y) = $f_{C1}(x,y)$ on a C1 part of the contour

$$\frac{\partial u(x,y)}{\partial x}=f_{C1}(x,y)\text{ on the other C2 part of the contour} \quad [3.175]$$

where $\vec{n}$ is the normal on $C2$ pointing outward towards D.

We write it to simplify the expressions:

$$\frac{d^2u(x,y)}{dx^2}+\frac{d^2u(x,y)}{dy^2}+f(x,y)=0 \qquad [3.176]$$

Using the *weighted residuals method*, the partial derivation equation is replaced with an integral equation.

For any v element of a complete basis of functions of two variables, if $u(x,y)$ confirms the following equation:

$$\iint_D v(x,y)\left(\frac{d^2u(x,y)}{dx^2}+\frac{d^2u(x,y)}{dy^2}+f(x,y)\right)dxdy=0 \qquad [3.177]$$

then, it is equivalent to the partial derivative equation.

We have:

$$\frac{d^2u(x,y)}{dx^2}+\frac{d^2u(x,y)}{dy^2}+f(x,y)=0 \Leftrightarrow$$

$$\iint_D v(x,y)\left(\frac{\partial^2u(x,y)}{\partial x^2}+\frac{\partial^2u(x,y)}{\partial y^2}+f(x,y)\right)dxdy=0 \qquad [3.178]$$

We carry out integration by parts on:

$$\iint_D v(x,y)\frac{\partial u(x,y)}{\partial x}dxdy=\int_{y1}^{y2}dy\int_{x1(y)}^{x2(y)}v(x,y)\frac{\partial u(x,y)}{\partial x}dx \qquad [3.179]$$

We put $a(x,y)=v(x,y)$ and $b'(x,y)=\dfrac{\partial u(x,y)}{\partial x}$ [3.180]

Thus, $a'(x,y)=\dfrac{\partial v(x,y)}{\partial x}$ et $b(x,y)=f(x,y)$ [3.181]

Hence, we get:

$$\iint_D v(x,y)\frac{\partial u(x,y)}{\partial x}dxdy = \int_{y1}^{y2} dy\left(\left[v(x,y)u(x,y)\right]_{x1(y)}^{x2(y)} - \int_{x1(y)}^{x2(y)} \frac{\partial v(x,y)}{\partial x}u(x,y)dx\right)^2 \quad [3.182]$$

$$\iint_D v(x,y)\frac{\partial u(x,y)}{\partial x}dxdy = -\iint_D u(x,y)\frac{\partial v(x,y)}{\partial x}dxdy + \int_{y1}^{y2} v\left[x2(y),y\right]u\left[x2(y),y\right]dy - \int_{y1}^{y2} v\left[x1(y),y\right]u\left[x1(y),y\right]dy \quad [3.183]$$

$$\iint_D v(x,y)\frac{\partial u(x,y)}{\partial x}dxdy = -\iint_D u(x,y)\frac{\partial v(x,y)}{\partial x}dxdy + \oint_C v(x,y)u(x,y)\cos\theta dl \quad [3.184]$$

Similarly, we get:

$$\iint_D v(x,y)\frac{\partial u(x,y)}{\partial y}dxdy = -\iint_D u(x,y)\frac{\partial v(x,y)}{\partial y}dxdy + \oint_C v(x,y)u(x,y)\sin\theta dl \quad [3.185]$$

Going to second derivatives:

$$\iint_D v(x,y)\frac{\partial^2 u(x,y)}{\partial^2 x}dxdy = -\iint_D \frac{\partial u(x,y)}{\partial x}\frac{\partial v(x,y)}{\partial x}dxdy + \oint_C v(x,y)\frac{\partial u(x,y)}{\partial x}\cos\theta dl \quad [3.186]$$

$$\iint_D v(x,y)\frac{\partial^2 u(x,y)}{\partial^2 y}dxdy = -\iint_D \frac{\partial u(x,y)}{\partial y}\frac{\partial v(x,y)}{\partial y}dxdy + \oint_C v(x,y)\frac{\partial u(x,y)}{\partial y}\sin\theta dl \quad [3.187]$$

Thus:

$$\iint_D v(x,y)\left(\frac{\partial^2 u(x,y)}{\partial x^2}+\frac{\partial^2 u(x,y)}{\partial y^2}\right)dxdy = -\iint_D\left(\frac{\partial u(x,y)}{\partial x}\frac{\partial v(x,y)}{\partial x}+\frac{\partial u(x,y)}{\partial y}\frac{\partial v(x,y)}{\partial y}\right)dxdy + \oint_C v(x,y)\frac{\partial u(x,y)}{\partial n}dl \quad [3.188]$$

Going back to the initial form, we get:

$$-\iint_D\left(\frac{\partial u(x,y)}{\partial x}\frac{\partial v(x,y)}{\partial x}+\frac{\partial u(x,y)}{\partial y}\frac{\partial v(x,y)}{\partial y}-v(x,y)f(x,y)\right)dxdy + \int_{C1} v(x,y)\frac{\partial u(x,y)}{\partial n}dl_1 + \int_{C2} v(x,y)u(x,y)dl_2 = 0 \quad [3.189]$$

We put: v(x,y)=0 in C1. Thus:

$$-\iint_D\left(\frac{\partial u(x,y)}{\partial x}\frac{\partial v(x,y)}{\partial x}+\frac{\partial u(x,y)}{\partial y}\frac{\partial v(x,y)}{\partial y}-v(x,y)f(x,y)\right)dxdy + \int_{C2} v(x,y)u(x,y)dl_2 = 0 \quad [3.190]$$

Now, we only have to choose the basis of the finite-dimensional subset E. This base is called $\{\varphi_j\}_{j=1}^{j=m}$.

To choose the basis, we will analyze the following conditions for φ_j functions:

– boundary conditions;

– they must be infinitely differentiable functions;

– derivatives must be square-integrable.

Discretization is carried out. We cut D set into pieces, thereby getting the following expression for the ith element:

$$\iint_D v_i(x,y)\left(\frac{\partial^2 u(x,y)}{\partial x^2}+\frac{\partial^2 u(x,y)}{\partial y^2}+f(x,y)\right)dxdy = 0 \quad [3.191]$$

For every element e, polynomial approximation is carried out on $u(x,y)$. Thus, $u(x,y)$ is substituted by:

$$u_e = \sum_j P_j(x,y).u_j(x,y). \qquad [3.192]$$

For the $v(x,y)$ function in an element e becomes:

$$v_e(x,y) = P_e(x,y)\delta u_e \qquad [3.193]$$

where δu_e are arbitrary variations of $u_e(x,y)$. $u_e(x,y)$ are values of $u(x,y)$ at mesh nodes.

$$\delta u_e \iint_D P_e(x,y)\left(\frac{\partial^2 \sum_j P_j(x,y).u_j(x,y)}{\partial x^2} + \frac{\partial^2 \sum_j P_j(x,y).u_j(x,y)}{\partial y^2} + f(x,y)\right)dxdy = 0 \qquad [3.194]$$

Going to the matrix form:

For every element, we define an elementary domain D_e.

$$\begin{cases} W_e = \iint_{D_e} \delta u\left(\frac{\partial^2 u(x,y)}{\partial x^2} + \frac{\partial^2 u(x,y)}{\partial y^2} + f(x,y)\right)dxdy \\ \sum_e W_e = 0 \end{cases} \qquad [3.195]$$

We put:

$$\begin{cases} \partial\vec{u}(x,y) = \left[u(x,y), \frac{\partial u(x,y)}{\partial x}, \frac{\partial u(x,y)}{\partial y}, \frac{\partial^2 u(x,y)}{\partial^2 x}, \frac{\partial^2 u(x,y)}{\partial^2 y}\right] \\ \partial\delta\vec{u}(x,y) = \left[\delta u(x,y), \frac{\partial \delta u(x,y)}{\partial x}, \frac{\partial \delta u(x,y)}{\partial y}, \frac{\partial^2 \delta u(x,y)}{\partial^2 x}, \frac{\partial^2 \delta u(x,y)}{\partial^2 y}\right] \end{cases} \qquad [3.196]$$

Thus, we can write:

$$W_e = -\iint_{D_e} \delta(\partial\vec{u}(x,y))[D]\partial\vec{u}(x,y)dxdy + \iint_{D_e} \delta u(x,y)f(x,y)dxdy + \oint_{C_e} \delta u(x,y)f_{C_e}(x,y)dl \qquad [3.197]$$

We put: $\begin{cases} u(x,y) - \vec{P_e}(x,y)\vec{u_e} \\ \delta u(x,y) - \vec{P_e}(x,y)\overrightarrow{\delta u_e} \end{cases}$ [3.198] with: $$\begin{cases} \overrightarrow{u_e} = \left[\ldots u_j \ldots \right] \\ \overrightarrow{P_e}(x,y) = \left[\ldots P_{ej}(x,y) \ldots \right] \end{cases} \quad [3.199]$$

We also put: $$\begin{cases} \partial \vec{u}(x,y) = \left[B(x,y) \right] \overrightarrow{u_e} \\ \partial \partial \vec{u}(x,y) = \left[B(x,y) \right] \overrightarrow{\delta u_e} \end{cases} \quad [3.200]$$

So, we get:

$$\begin{aligned} W_e = & -\iint_{D_e} {}^t\overrightarrow{\delta u_e}\, {}^t\left[B(x,y) \right]\left[D \right]\left[B(x,y) \right] \overrightarrow{u_e} dxdy \\ & + \iint_{D_e} {}^t\overrightarrow{\delta u_e} \overrightarrow{P_e}(x,y) f(x,y) dxdy \\ & + \oint_{C_e} \overrightarrow{\delta u_e} \overrightarrow{P_e}(x,y) f_{C_e}(x,y) dl \end{aligned} \quad [3.201]$$

Hence, we get a matrix system for the element:

$$\begin{cases} W_e = -{}^t\overrightarrow{\delta u_e}\left(\left[k_e \right] \overrightarrow{u_e} - \overrightarrow{F_e} \right) \\ \left[k_e \right] = \iint_{D_e} {}^t\left[B(x,y) \right]\left[D \right]\left[B(x,y) \right] dxdy \\ \overrightarrow{F_e} = \iint_{D_e} \overrightarrow{P_e}(x,y) f(x,y) dxdy + \oint_{C_e} \overrightarrow{P_e}(x,y) f_{C_e}(x,y) dl \end{cases} \quad [3.202]$$

Since we are working in two dimensions, the $\overrightarrow{P_e}(x,y)$ expression is very complex, as it depends on both the form of the element and the position of nodes. Therefore, we will work on $u(\xi,n)$ instead of working with $u(x,y)$.

We put $$u(\xi,n) = \sum_j P_{ej}(\xi,n) u_j = {}^t\overrightarrow{P_e}(\xi,n).\overrightarrow{u_e} \quad [3.203]$$

The Jacobian transformation allows for going from the reference element to the real element:

$$\tau\begin{cases} x = x(\xi,\eta) \\ y = y(\xi,\eta) \end{cases} \text{ and } [J] = \begin{pmatrix} \partial x/\partial\xi & \partial y/\partial\xi \\ \partial x/\partial\eta & \partial y/\partial\eta \end{pmatrix} \quad [3.204]$$

where:

$$du = \frac{\partial u}{\partial x}dx + \frac{\partial u}{\partial y}dy = \frac{\partial u}{\partial x}\left(\frac{\partial x}{\partial\xi}d\xi + \frac{\partial x}{\partial\eta}d\eta\right) + \frac{\partial u}{\partial y}\left(\frac{\partial y}{\partial\xi}d\xi + \frac{\partial y}{\partial\eta}d\eta\right) \quad [3.205]$$

$$\text{and } \begin{pmatrix} \dfrac{\partial u}{\partial\xi} \\ \dfrac{\partial u}{\partial\eta} \end{pmatrix} = \begin{pmatrix} \dfrac{\partial x}{\partial\xi} & \dfrac{\partial y}{\partial\xi} \\ \dfrac{\partial x}{\partial\eta} & \dfrac{\partial y}{\partial\eta} \end{pmatrix}\begin{pmatrix} \dfrac{\partial u}{\partial x} \\ \dfrac{\partial u}{\partial y} \end{pmatrix} \quad [3.206]$$

Now, we can write:

$$\overrightarrow{\partial u}(\xi,\eta) = [J]\overrightarrow{\partial u}(x,y) \text{ and } \delta\overrightarrow{\partial u}(\xi,\eta) = [J]\delta\overrightarrow{\partial u}(x,y) \quad [3.207]$$

Using these notations, we get:

$$\overrightarrow{\partial u}(\xi,\eta) = [J][B(x,y)]\overrightarrow{u_e}(x,y) = [B(\xi,\eta)]\overrightarrow{u_e}(x,y) \quad [3.208]$$

$$\text{and: } [B(x,y)] = [J]^{-1}[B(\xi,\eta)] \quad [3.209]$$

We shift to a domain of the reference element D_r, C_r edge.

$$W_e = -{}^t\overrightarrow{\delta u_e}\left[\begin{array}{l} \iint_{D_r} {}^t[B(\xi,\eta)]\,{}^t[J]^{-1}[D][J]^{-1}[B(\xi,\eta)]\det[J]\overrightarrow{u_e}d\xi d\eta - \\ \iint_{D_r} \overrightarrow{P_e}(\xi,\eta)f(\xi,\eta)\det[J]d\xi d\eta - \oint_{C_r} \overrightarrow{P_e}(\xi,\eta)f_C(\xi,\eta)J_r dl_r \end{array}\right] \quad [3.210]$$

We put: $dl = J_r dl_r$

$$\begin{cases} W_e = -{}^t\delta\overrightarrow{u_e}\left([k_e]\overrightarrow{u_e} - \overrightarrow{F_e}\right) \\ [k_e] = \iint\limits_{D_r} {}^t[B(\xi,\eta)]\,{}^t[J]^{-1}[D][J]^{-1}[B(\xi,\eta)]\det[J]\overrightarrow{u_e}d\xi d\eta \\ \overrightarrow{F_e} = \iint\limits_{D_r} \overrightarrow{P_e}(\xi,\eta)f(\xi,\eta)\det[J]d\xi d\eta + \oint\limits_{C_r} \overrightarrow{P_e}(\xi,\eta)f_C(\xi,\eta)J_r dl_r \end{cases} \quad [3.211]$$

3.9. Finite elements, under MATLAB (NPN)

3.9.1. *Structure definition*

The finite element method is one of the mathematical tools applied, used in numerical analysis; it is the most suitable for numerically solving partial differential equations, here for complex geometries. First applied to mechanic structure problems, it is today applied to most practical problems that are linear or nonlinear, stationary or time-dependent, defined in a uni-, two- or three-dimensional domain, in order to represent the behavior of different physical as well as mechanical, thermodynamic, acoustic or electronic systems.

More specifically, it makes it possible to, for example, numerically calculate object behaviors, even complex ones, provided that they are "continuous" and described by a linear partial differential equation: *motion of a cord shaken by one of its ends, behavior of a rapidly moving fluid on an obstacle, deformation of a metallic structure, etc.*

We are talking about a *discrete mathematical algorithm* that makes it possible to find an approximate solution to a partial differential equation (PDE) in a domain having edge and/or internal conditions.

First, we need to have an approximate solution to a problem, wherein solutions to that problem verify existence conditions weaker than those of initial problem solutions and where *discretization** can help find an approximate solution. As with many other numerical methods, it is important to know the *quality of discretization*: existence and uniqueness of the solution, convergence, stability, and of course, the difference in error between our discrete solution and the unique solution to the initial problem.

The finite element method discretely solves a PDE for which we want a "sufficiently" reliable approximate solution. Generally, the PDE concerns a certain function, u, defined in a domain. It has *edge conditions*, ensuring that the solution exists and is unique. *Discretization** requires defining an appropriate space of test function, in which space solutions for variational formulation of the equation are accurate. This requires that a *mesh* of the fragmented domain is defined: *finite elements*. These fragments can, for example, be of any form but must pave the concerned space. Usually, finite elements are triangular or rectangular. In this context and giving special emphasis to the domain edge problem, we get an algebraic formulation: discretization of the initial problem. The solution to this algebraic problem, if it exists and is unique, gives an approximate solution.

Once the solution is found, it is important to accurately estimate the error concerning discretization and prove that this method thus written converges, which means that the error tends towards 0 if the fineness of the mesh also tends towards 0.

Discretization and mesh

In most cases, equations are impossible to solve on a large domain (even if integral). Thus, we carry out an approximation to facilitate integral calculation:

For example, it is easier to calculate

$$\int \sin(x)\sqrt{xdx} \quad [3.212]$$

If we approximate sin(x) using

$$x+\frac{x^3}{3}+\ldots! \quad [3.213]$$

However, as shown, approximation is valid around a point (above, for x = 0). If the problem is solved by *discretizing* an unknown function, we divide the definition of the equation into *several fragments (finite elements)* that must form cobbles *in the* concerned *space*: the peaks of the fragments form a *mesh*. Owing to this, we can approximate the unknown function in each peak (or element node) using a simpler approximate function which will lead to a solution to the problem.

In 2D, we can analogize that a hammock (rope meshing) is the approximation of a sheet. In 3D, we can assimilate discretization of a volume using crystallography concepts, which reflect the fact that matter is not continuous.

The form of the mesh is a vital element in this method as it determines the points on which we calculate the function. If the mesh form is not adapted (bad form or not dense enough), we will not have an accurate solution. For example, if we try to discretize mountainous terrain to learn about that terrain, but do not choose the mountain peaks as mesh nodes and the valleys as the lowest points, we will end up not having a correct representation of this terrain and any study on this representation will be inaccurate. Thus, it is not necessary that the mesh be regular, and we tend to tighten the places of interest on the mesh (for instance, places where we think the solution will vary a lot).

We have an integral solution for each element, involving a k matrix, the coefficients of which and the fi of various nodes (generally, the concerned node and the nodes around it) are determined.

3.9.2. *Assembly*

When solving an equation system for all nodes, it is not enough to consider only one insulated element. In a finite element mesh, a node generally belongs to several elements and the load on each one of the elements results in the f value in the node.

The value of the function is calculated element by element; yet, as peaks are common to several elements, all local systems must be assembled into one large system containing the equations of all nodes.

To construct this system, we will analyze the initial differential equation in each element, which means that we will have equations in each element peak. The equation on each peak depends on the values taken from the neighboring peaks and the partial derivative.

When we put equations of each peak together, we get a system where the unknowns are values of the functions on mesh nodes (peaks) (generally, n unknowns and n-x equations, x being the size of E).

3.9.3. *Solving: pros and cons of the finite element method*

This is when we apply boundary conditions that lead to another new equation, making it possible to solve the system.

We determine the fi, f values at each node, enabling us to trace the function using approximation.

When we have the fi, all we have to do is carry out linear interpolation of these fi to obtain f (approximate values) at each point.

If we do not tighten the mesh enough (in order to manually calculate in a relatively shorter time), we will have huge gaps between the real solution and the approximated one. However, if we want only values of the function at specific points in the domain, then this simple mesh will be enough. The power of computers now available makes it possible to use highly tightened meshes, which require heavy calculations, and to have highly acceptable results (see "Method error").

An important advantage of finite elements is the flexibility in mesh choices: element form, position and number of nodes, etc., as long as we tighten the mesh on specific points (or on peaks). Thus, a contour of some form can be approached from as close as we want, using a set of polygons. Using a non-uniform mesh is not a problem, given that integral calculations are carried out simultaneously on an element. So, we can tighten or loosen the mesh at desired places to increase precision or eliminate useless calculations. The only disadvantage of this method is the complex programming or computation involved. The time spent decrypting, creating or even adapting a program for finite element solving is huge, and must be adaptive to various boundary conditions in order to efficiently use it.

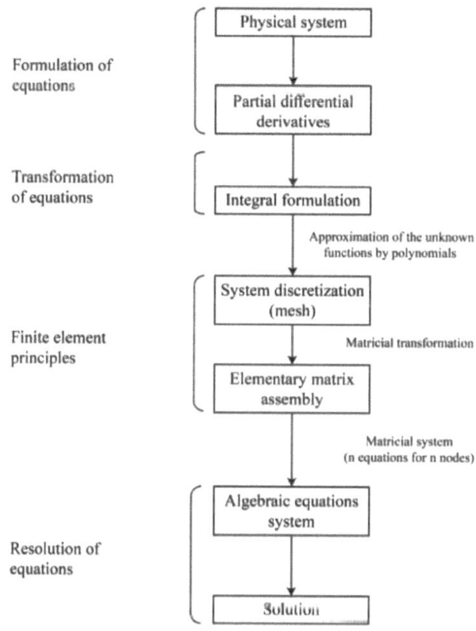

Finite elements organizational chart

Figure 3.74. *Flow chart of a DDM algorithm resolution*

4

Background Noise in Micro- and Nanoelectronics

SUMMARY.– Noise is a universal phenomenon. Generally, microelectronics and, of course, nanoelectronics are domains where even autonomous noise sources can be dependent on each other due to their proximity, for example. Here, we will introduce concepts such as the correlation of random variables and noise spectral density. We will present the paradigm of transfer impedance. Although calculations at the unidimensional level have already been developed, there are only a few in 2D and nothing considerable in 3D. Speed fluctuations, which can help quantify certain noises using master equations such as the Boltzmann equation, can be taken into account in statistical methods like Monte Carlo.

4.1. Introduction

Laboratories work in the domain of background noise in semiconductors and devices, and thus on the fluctuation of physical or electrical quantities and *ad hoc* mathematical tools, thereby building a comprehensive and cooperative program via a dialogue between researchers, students, postgraduates, devices and information. This covers fundamental studies as well as their applications in modern submicronic devices (submicronic (SiGeC) silicon transistors, ultimate CMOS, high-speed III–V devices (HEMT: High-Electron Mobility Transistor, Heterojunction Bipolar Transistor) and optoelectronic devices (lasers, photodiodes)).

In electronic chips, semiconductor components are immersed in a more or less noisy environment. From a microscopic point of view, fluctuations of physical and electronic quantities can be apprehended, for instance, by using the Boltzmann equation, numerically or via Monte Carlo methods, and by calculating, for example, the speed fluctuation correlation function and its Fourier transform, which will give the noise spectral density.

It was Shockley [SHO 66], the inventor of the bipolar transistor (1948), who initiated these studies. Van der Ziel [VAN 86], [VAN 70] and Van Vliet [VAN 75], as well as other eminent researchers, contributed to the studies.

Currently, related numerical methods, such as the very promising impedance transfer method, which can lead to some results, but, most of the time, it will lead to a one-dimensional size.

4.2. Fluctuations and noise [JIN 94, LAN 08]

4.2.1. *Recap on Dirac distribution*

The Dirac function is defined for a real variable x using:

$$f(x_0) = \int_x f(x)\delta(x - x_0)\, dx \tag{4.1}$$

where f is some function and Δx is the space surrounding the point x_0.

Taking f (x) = 1, we see that $\delta(x - x_0)$ is zero everywhere except x = x_0 where it is infinite:

$$\int_\Delta \delta(x - x_0)\, dx = 1 \tag{4.2}$$

$\int_{\Delta x} \delta(x - x_0)$ can also be represented by the limitation of some functions, including the pulse of the length ε and the height 1/ε, centered about x_0, or by a Gaussian:

$$\delta(x - x_0) = lim_{\sigma \to 0} \frac{1}{\sigma\sqrt{2\pi}}^{\frac{-(x-x_0)^2}{2\sigma^2}} \tag{4.3}$$

H (x – x_0) being the Heaviside step function, we get:

$$\text{H}(\text{x} - \text{x}_0) = \begin{cases} 1 \text{ } if\ x \geq x_0 \\ o \text{ } if\ x < x_0 \end{cases} \tag{4.4}$$

$$\delta(x) = \frac{dH\prime x)}{dx} \tag{4.5}$$

$$1 = \int_{-\infty}^{\infty} e^{2i\pi vx}\, \delta(x) dx \tag{4.6}$$

$$\delta(x) = \int_{-\infty}^{\infty} e^{2i\pi vx}\, dv \tag{4.7}$$

For several space variables, 3, for example:

d^3r=dx dy dz, in Cartesian coordinates

$$\delta(\vec{r} - \vec{r_0}) = \delta(x - x_0)\delta(y - y_0)\delta(z - z) \quad [4.8]$$

In addition, for a function hs:

$$h(\vec{r})\delta(\vec{r} - \vec{r_0}) = h(\vec{r_0})\delta(\vec{r} - \vec{r_0}) \quad [4.9]$$

4.2.2. *Random functions, correlation and spectral densities of fluctuations*

4.2.2.1. *Random functions*

A random variable is a real or complex number, the value of which depends on chance (that is, *a priori* unpredictable), meaning that it depends on a "proof". A random function is one for which the value is constantly dependent on chance. A simple random function depends on one or more random variables. To find out the value of the function at a specific moment t, there is a random variable x (f) that can be described by its probability density p (x, t) or, briefer still, by certain moments: $<X^1(t)>$, $<X^2(t)>$, etc. (average values are also called expectations and marked E, thus E{X (t)}, E{$X^2(t)$}, …).

It will also be interesting to look at properties at two given instants: a moment that plays a specifically important role is covariance or correlation:

$$\Gamma(X(t_1, t_2)) = < X(t_1)X^*(t_2)> \quad [4.10]$$

$X^*(t_2)$ is the complex conjugate of $X(t_2)$.

4.2.2.2. *Second-order stationary random functions*

In numerous cases, we deal with at least second-order random functions. These functions have the following properties:

a) They are first-order stationary:

$$<X(t)> = m \quad [4.11]$$

b) Let us use second-order stationarity

$$\Gamma_X(\tau) = < X(t) X^*(t - \tau) > \quad [4.12]$$

For $\tau = o$, $\Gamma_X(0) = < |X(t)^2|>$

From the stationarity, we directly get: $\Gamma_X(\tau) = \Gamma^*_X(-\tau)$ [4.13]

4.2.2.3. *Wiener–Khinchin theorem*

If the signal can be written as a Fourier–Stieltjes transform, an integration of complex exponentials over the elementary increase dx(ν) can be expressed as:

$$X(t) = \int_{-\infty}^{\infty} e^{2i\pi\nu t}\, dx(\nu), \quad [4.14]$$

Taking into account [4.11] and [4.14], the average, which is constant, can be written as:

$$m = \int_{-\infty}^{\infty} e^{2i\pi\nu t} < dx(\nu) > \quad [4.15]$$

from which, we get:

$$< dx(\nu) > = m.\delta(\nu)d\nu \quad [4.16]$$

Thus, $< dx(\nu) >$ is independent of ν and non-zero only when ν = 0.

Now, if we take second-order stationarity into account, that is, if we substituted X(t) and X* (t - τ) with their values:

$$\Gamma_X(\tau) = < \int_{-\infty}^{+\infty} e^{2i\pi\nu t}\, dx(\nu). \int_{-\infty}^{+\infty} e^{-2i\pi(t-\tau)}\, dx^*(\nu') > \quad [4.17]$$

$$\Gamma_X(\tau) = \int_{-\infty}^{+\infty} \int_{-\infty}^{+\infty} < dx(\nu)dx^*(\nu') > e^{2i\pi(\nu')\tau} e^{2i\pi(\nu-\nu')\tau} \quad [4.18]$$

$\Gamma_X(\tau)$ must be independent of t, and thus all ν ≠ν terms are zero.

Then, from integral [4.18], we can write:

$$< dx(\nu)\, dx^* (\nu') > = \delta(\nu - \nu')\gamma_X(\nu)d\nu d\nu', \quad [4.19]$$

Thus, there is no correlation between different frequencies for second-order stationary functions. The γ_x (ν) quantity is called the spectral density of fluctuation of X. It is observed on the whole frequency spectrum, positive and negative. In general, we only take positive frequencies, and the measurable physical quantity S_X (f) is defined by:

$$S_X (f) = \gamma_x (\nu=f) + \gamma_x ((\nu= -f) \quad [4.20]$$

$$\Gamma_X(\tau) = \int_{-\infty}^{+\infty} \int_{-\infty}^{+\infty} \delta(\nu - \nu')\gamma_X(\nu) e^{2i\pi\nu'\tau}\, e^{2i\pi(\nu-\nu')\tau}\, d\nu\, d\nu' \quad [4.21]$$

After integrating on ν',

$$\Gamma_X(\tau) \;=\; \int_{-\infty}^{+\infty} e^{2i\pi\nu\tau}\, \gamma_X(\nu)\, d\nu \qquad [4.22]$$

$\Gamma_x(\tau)$ is the Fourier transform of the spectral density:

$$\gamma_X(\nu) = \int_{-\infty}^{+\infty} e^{-2i\pi\nu\tau}\, \Gamma_X(\tau)\, d\tau \qquad [4.23]$$

Equations [4.12] and [4.23] relate to the Wiener–Khinchin theorem.

4.2.2.4. *Properties of spectral density of fluctuation*

a) Complex conjugate of equation [4.23]:

taking [4.23] into account and adding $\theta = -\tau$, we get:

$$\gamma_X^*(\nu') = -\int_{-\infty}^{+\infty} \Gamma_X^*(-\theta) e^{-2i\pi\nu\theta}\, d\theta = \int_{-\infty}^{+\infty} \Gamma_X(\theta) e^{-2i\pi\nu\theta}\, d\theta = \gamma_X(\nu) \qquad [4.24]$$

Hence, $\gamma_X^*(\nu)$ is a real quantity, which is positive.

b) White noise: the noise is said to be white if $\gamma_X(\nu)$ is constant, irrespective of ν:

Then, equation [4.22] becomes:

$$\Gamma_X(\tau) = \int_{-\infty}^{+\infty} A\, e^{-2i\pi\nu t}\, d\tau = A\, \delta\tau \qquad [4.25]$$

When a noise is white, it means that successive events are not correlated. In practice, correlation is non-zero at small time intervals τ_0.

τ_0 is the relaxation time of the system, and we get: $\Gamma_X(\tau) = 0$ if $\tau > \tau_0$, that is, $\gamma_X(\nu)$ is constant when $\nu < 1/\tau_0$.

4.3. Noise calculation using the Langevin method

4.3.1. *Principle of the Langevin method [THO 74]*

In a very general scope, the Langevin method, which is applied to electronic noise, involves replacing the real device with a no-noise device in series with a noise source.

If η is the noise source, then the linear Langevin operator applied to the random variable is the said noise source, that is,

$$\hat{L}X(t) = \eta(t) \qquad [4.26]$$

If the average value of X(t) is zero, then we get:

$$\hat{L} < X(t) >=< \eta(t) >= 0 \qquad [4.27]$$

In addition,

$$X(t) = \int_{-\infty}^{\infty} e^{2i\pi\nu t}\, dx(\nu) \qquad [4.28]$$

$$\eta(t) = \int_{-\infty}^{\infty} e^{2i\pi\nu t}\, d\varsigma(\nu) \qquad [4.29]$$

where $e^{2i\pi\nu t}$ is an eigenvector of $\hat{L}$, with Z(ν) being its eigenvalue. Thus, we can write:

$$\hat{L}\, e^{2i\pi\nu t} = Z(\nu).\, e^{2i\pi\nu t} \qquad [4.30]$$

Therefore, the noise spectral densities are connected via:

$$S_\varsigma (f) = |Z'(\nu, f)^2| S_x (f) \qquad [4.31]$$

We see that the phase does not interfere here with regard to second-order properties.

4.3.2. *Example: generation–recombination noise*

In a semiconductor bar, we can generate ΔN number of new carriers by lighting, for instance. If τ is the carrier lifetime, we can write:

$$\frac{d\Delta n}{dt} = -\frac{\Delta n}{\tau} \qquad [4.32]$$

If we focus on fluctuations:

$$\frac{d\Delta n}{dt} = -\frac{\Delta n}{\tau} + N(t) \qquad [4.33]$$

If the source is harmonic:

$$N(t) = N_0 e^{j\omega t} \qquad [4.34]$$

Δn is also harmonic (see linearity): $\Delta n(t) = \Delta N e^{j\omega t}$ [4.35]

Then, we get:

$$S_N(f) = S_{\Delta n}(f) \left|\frac{1}{\tau} + j\omega\right|^2 \quad [4.36]$$

$$S_{\Delta N}(f) = S_\eta(f) \frac{\tau^2}{1+\omega^2\tau^2} \quad [4.37]$$

If it is white noise,

$$< \Delta N^2 > = \int_0^\infty S_{\Delta N}(f)df \text{ ^}= S_{S_\eta} \tau^2 \int_0^\infty \frac{1}{1+4\pi^2 f^2 \tau^2} df = S_\eta \frac{\tau}{4} \quad [4.38]$$

Thus:

$$S_{\Delta N}(f) = 4 < \Delta N^2 > \frac{\tau}{1+\omega^2\tau^2} \quad [4.39]$$

We know the spectrum if $< \Delta N^2 >$ is white. We then use a master equation that regulates microscopic phenomena.

4.3.3. *Quantities associated with noise in electricity*

Let us take the Langevin theorem again:

$$S_V(f) = |Z(f)|^2 S_I(f) \quad [4.40]$$

$$Z(f) = \frac{\Delta V}{\Delta I} \quad [4.41]$$

ΔV is complex voltage (see sinusoidal regime), superimposed at the quiescent point, continuous and induced by ΔI.

We can then define an equivalent noise temperature, T_n, as:

$$S_V(f) = 4k T_n(f) \text{ Re}(Z(f)) \quad [4.42]$$

or

$$S_I(f) = 4k T_n(f) \text{ Re}(Y(f)) \quad [4.43]$$

with

$$Z(f) - 1/Y(f). \quad [4.44]$$

Finally, we can define an equivalent noise current, I_n, as:

$$S_I(f) = 2q.I_n(f) \qquad [4.45]$$

Thus, spectral densities are linked to the square modulus of the complex impedance. This concerns macroscopic fluctuations. If the source term is known, the fluctuation of the studied quantity can be deduced. It is also noted that even if the source has white noise, a normal assumption, the studied variable does not generally have white noise. Z depends on the frequency.

4.3.3.1. *Observations*

1) To know one of the two terms, I (f) or S (f), we use microscopic conditions via master equations (Boltzmann, Fokker–Planck equations, etc.),

2) We have seen that it is absolutely necessary for the operator to be linear in order for equation [4.26] to be applied; else, the Langevin description cannot be applied.

It is only when L is linear that we get a noise source with an average value of zero, and thus find the no-noise system equation.

If the system is not linear, equation [4.26] can be linearized because, in general, we focus on the small fluctuations around the mean value. Therefore, we can use $X(t) = X_o + \Delta X(t)$ and linearize equation [4.26], given that $< Xo> = 0$.

4.3.4. *Example application: thermal noise of a resistor*

Let us take a circuit (L, R). We find the spectral density of the noise current in the circuit through which a zero average current flows. For this, we insert a source v(t) in the perfect circuit and the circuit obeys the equation:

$$L\frac{di}{dt} + Ri = v(t) \qquad [4.46]$$

If the source is harmonic, we get:

$$v=V_m.e^{j\omega t} \qquad [4.47a]$$

and

$$i=I_m.e^{j\omega t} \qquad [4.47b]$$

The result is well known:

$$V_m = Z\ I_m \qquad [4.48a]$$

with

$$Z=R+jl\omega \quad [4.48b]$$

Thus:

$$S_V(f)=|Z|^2 S_I(f) \quad [4.49a]$$

$$S_I(f)=\frac{S_V(f)}{R^2+L^2\omega^2} \quad [4.49b]$$

Now, let us consider the microscopic aspect and calculate $<i^2>$.

The average energy stored inside the coil is: $\frac{1}{2}.Li^2$ [4.50a]

In 1D, we can write:

$$\tfrac{1}{2}.Li^2 = \tfrac{1}{2}.kT \quad [4.50b]$$

Yet:

$$< i^2 >= \int_0^{\infty} S_i(f)\ df\ ) = \int_0^{\infty} \frac{df}{R^2+4\pi^2L^2f^2} \quad [4.51a]$$

$$=\frac{S_V}{4RL} \quad [4.51b]$$

Therefore:

$$S_V = 4kRT \text{ (Nyquist theorem)} \quad [4.52]$$

$$S_I = 4kT\frac{R}{R^2+L^2\omega^2} \quad [4.53]$$

4.3.5. *Forms of Langevin equations*

Langevin equations often involve several coupled equations for X_1 (t), X_2(t) variables, the average value of which satisfies the relationship of the form:

$$\frac{\partial <x_i(t)>_{cond}}{\partial t} + \sum_i M_{ij} < x_i(t) >_{cond} = 0 \quad [4.54]$$

$< x_i(t) >_{cond}$ is a conditional mean that satisfies the given initial conditions. This is the case, for example, with electric circuits where initial conditions must be set; these equations are valid around the polarization point for all nonlinear systems. If derivatives of an order higher than 1 appear, the order of the system is to be reduced; for example, for a second-order system, we construct a new random variable, that is, the first derivative of the initial variable.

It can be shown that these phenomenological equations are equivalent to a Langevin description, that is, the set of unconditional regression equations.

The related Langevin description is:

$$\frac{\partial < x_i(t) >}{\partial t} + \sum_j M_{ij} < x_j(t) > = \varsigma(t) \qquad [4.55]$$

given that the source terms are mean zero and have white noise.

4.3.6. *Example: electron injection in the p area of a PN junction*

In an ohmic regime, the transport equation without generation–recombination ($-\Delta n/\tau$) can be written as:

$$\frac{\partial < \Delta n \ (\vec{r},t) >_{cond}}{\partial t} = -\frac{< \Delta n \ (\vec{r},t) >_{cond}}{\tau} + \frac{1}{e} \nabla < \Delta \vec{j}_n(\vec{r},t) >_{cond} \qquad [4.56]$$

Moreover:

$$< \Delta j_n(\vec{r},t) >_{cond} = -e\,\mu_n < \Delta n \ (\vec{r},t) >_{cond} \vec{E} + e\,D_n < \Delta n \ (\vec{r},t) >_{cond} \qquad [4.57]$$

$$\frac{\partial \Delta n \ (\vec{r},t)}{\partial t} = -\frac{\Delta n \ (\vec{r},t)}{\tau} + \frac{1}{e} \vec{\nabla} . \Delta \vec{j}_n(\vec{r},t) + \varsigma_n(\vec{r},t) \qquad [4.58]$$

$$\Delta j_n(\vec{r},t) = -e\,\mu_n \Delta n \ (\vec{r},t) \ \vec{E} + e D_n < \Delta n \ (\vec{r},t) >_{cond} \vec{\eta}_n(\vec{r},t) \qquad [4.59]$$

Analogously, for holes:

$$\frac{\partial \Delta p \ (\vec{r},t)}{\partial t} = -\frac{\Delta p \ (\vec{r},t)}{\tau} - \frac{1}{e} \vec{\nabla} . \Delta \vec{j}_{np}(\vec{r},t) + \varsigma_p(\vec{r},t) \qquad [4.60]$$

$$\Delta j_p(\vec{r},t) = +e\,\mu_p \Delta p \ (\vec{r},t) \ \vec{E} - e\,D_p \vec{\nabla} . \Delta n \ (\vec{r},t) >_{cond} + e\vec{\eta}_p(\vec{r},t) \qquad [4.61]$$

with ς_n *and* ς_p being noise sources.

4.4. Noise calculation using the impedance field method

4.4.1. *Principle of the method [NOU 91, VAN 75]*

The impedance field method was introduced by W. Shockley *et al.* [SHO 66]. This method differs from the Langevin method in the way it links the macroscopic quantity to elementary microscopic fluctuations. The spectral density of voltage fluctuations can be brought down to two basic calculations: the spectral density of voltage fluctuation and elementary fluctuations.

4.4.2. *Impedance fields*

Let us consider a polarized specimen between electrode N and the grounding. If we inject a little current ΔI, flowing out of N, at a coordinate point $\vec{r}$ in the volume, the potential of N varies by $\Delta V(N, \vec{r})$.

Each noise course at the coordinate point $\vec{r}$ in the substrate (and also in the devices) produces a noise in another point of the volume and the surface (and electrodes).

The induced noise depends on its sources and its way of propagation via the impedance field. In order to study the noise between two contacts on the surface or in the substrate, we inject a little current (noise) δI (complex number), at frequency f, at a coordinate point $\vec{r}$ (+/- $\overrightarrow{dr}$) in the substrate, superimposed on a continuous current between the contacts. This little current produces a potential V(f) in another contact. Thus, we can write:

$$\Delta V\,(r+dr) = Z\,(r+dr)\,\delta I \qquad [4.62]$$

Z(r; f) is an impedance.

The resulting difference in voltage is:

$$\delta V = \Delta V\,(r+dr) - \Delta V\,(r) = \vec{\nabla} Z.dr.\,\Delta I \qquad [4.63]$$

where Green's function $\vec{\nabla} Z$ (ΔI=1) [4.63] is called *the impedance field.*

δIdr can be described by the current density δj(r) in an infinitesimal volume dr^3:

$$\delta I.dr = \delta j(r).d^3r \qquad [4.64]$$

where δj is the current density ($A.m^{-2}$).

By substituting [4.64] into [4.63], we get:

$$\delta V_f (r) = (Z(r,f) - Z(r\text{-}dr,f)).\delta I = \vec{\nabla} Z(r,f).\overrightarrow{dr}.\ \delta I \qquad [4.65]$$

$$\delta V_f = \vec{\nabla} Z(r,f).\delta j(r).\ dr^3 \qquad [4.66]$$

Equation [4.67] means that a current fluctuation in the dr^3 volume produces a voltage difference between two other contacts. Integrating on the considered domain, the total voltage produced by these currents is:

$$\Delta V_f = \int \vec{\nabla} Z(r,f).\ \delta\vec{j}(r).\ dr^3 \qquad [4.67]$$

In order to explicitly calculate the noise density of $S_V(f)$, we project components, for example Cartesian, in the x, y and z directions; in a direction, we get:

$$\Delta V_f = \sum_\alpha \left[\int \frac{\partial Z(r,f)}{\partial \alpha} \vec{\nabla} Z.\delta j_\alpha(r).\,dr^3\right] \qquad [4.68]$$

Thus, the voltage noise spectral density at the junctions becomes

$$S_v(f) = 2\int_{-\infty}^{\infty} \Gamma_{\delta V(}N,\vec{r},t).\,e^{2\pi i f t}dt \qquad [4.69]$$

where $\Gamma_{\delta V(}N,\vec{r},\tau)$. is the correlation function:

$$\Gamma_{\delta V} = \left(\delta V_f\ (t).\,\delta V_f^*(t+\tau)\right) \qquad [4.70]$$

The density becomes:

$$S_v(f) = \sum_\alpha \sum_\beta \iint \frac{\partial Z'(r,f)}{\partial \alpha}.\frac{\partial Z^*(r',''f)}{\partial \beta}.2.\int_{-\infty}^{\infty} \overline{\delta j_\alpha\ \ \delta j_\beta^*(r',t+\tau)}\,e^{2\pi f i \tau}d\tau.\,dr^3dr'^3$$

[4.71]

$$S_v(f) = \sum_\alpha \sum_\beta \iint \frac{\partial Z'(r,f)}{\partial \alpha}.\frac{\partial Z'(r',f)}{\partial \beta}.\,S_{j\alpha\beta}(r,r'',f)dr^3dr'^3 \qquad [4.72]$$

$$S_v(f) = \iint \vec{\nabla} Z'(r,f)\cdot Sj(r,r'',f).'\,\vec{\nabla} Z^*\ (r'',f)\quad dr^3dr'^3 \qquad [4.73]$$

If K is a noise source:

$$S_{j\alpha\beta}(r,r'',f) = K_{\alpha\beta}(r,f).\delta\,(r'-r) \qquad [4.74]$$

Equation [4.74] simplifies into:

$$S_v(f) = \int \vec{\nabla} Z \ (r,f) \cdot K(r,f). \ \ \vec{\nabla} Z^*(r',f) \qquad [4.75]$$

4.4.3. *Calculation of the impedance field*

Calculating the impedance field is easy in the case of homogeneous samples. However, in the case of non homogeneous samples, in the space charge regime, for instance, some precautions must be taken.

a) We determine the working regime, that is, the polarization point, by:

- solving local equations (Poisson's equation, electric neutrality, phenomenological expression of current density, etc.);

- taking boundary conditions into account, that is, for electrodes, to determine integration constants in the preceding differential equations;

Thus, we get the local mean field $E_0(t)$ based on the average current density j_o.

The average voltage at junctions in the sample is:

$$V_0 = -\int_{N_0}^{N} \vec{E}_0\,(\vec{r}).d\vec{r} \qquad [4.76]$$

where No and N are the electrodes. The average current is:

$$I_0 = \iint \vec{J_0}.\vec{D_S} \qquad [4.77]$$

where S is a slice of the sample, for instance the surface of an electrode.

b) We calculate the fluctuations of the electric field resulting from introducing the current ΔI at the coordinate point $\vec{r}$. The $\Delta\vec{E}\,(\bar{r},\vec{r}')$ field is obtained as follows:

- In local equations:

$$\vec{E}\,(\vec{r}\,') = \vec{E_0}(\vec{r}\,') + \Delta\vec{E}\,(\bar{r},\vec{r}\,') \text{ and } \vec{j}(\vec{r}\,') = \vec{j_0}(\vec{r}\,') + \Delta\vec{j}\,(\bar{r},\vec{r}\,') \qquad [4.78]$$

The Δ is considered as disturbances with regard to E_o and I_0; these zero-order terms give the equations for mean values, while the first-order terms in ΔE and ΔI give linear equations. The local Langevin equations thus obtained are solved and the integration constants are determined by imposing the same boundary conditions on $\Delta\vec{E}$ as those applied on $\vec{E_0}$. For $\Delta\vec{E}$, these conditions are at the level of electrode N and the $\bar{r}$ point, and at N and No for $\vec{E_0}$ Thus, we get $\Delta\vec{E}\,(\bar{r},\vec{r}\,')$.

c) We then deduce the impedance gradient.

4.4.4. *Example application: noise in the ohmic space charge regime (simple injection)*

The sample is assumed to be parallelepiped; here, we deal with a longitudinal unidimensional case. The space charge is not negligible and can make the characteristics of the device nonlinear.

We distinguish between the two different regimes:

a) when the space charge of the injected carriers is too weak to significantly affect the volume's conductivity, we get the ohmic regime, where the mean current is proportional to the voltage at the junctions of the sample $I_o \propto V_o$:

$$j = -\,q\mu n E + \varepsilon \frac{\partial E}{\partial t} \quad [4.79]$$

$$\frac{\partial E}{\partial x} = \frac{-q}{\varepsilon}\,(n - N_D) \quad [4.80]$$

$$j_n(\vec{r},t) \;=\; e\,\mu_n E(N_{D+}\frac{\varepsilon}{e}\frac{\partial E}{\partial t}) + \varepsilon\frac{\partial E}{\partial t} \quad [4.81]$$

$(q = -e, >0)$

As for fluctuations:

$$E = E_0 + \Delta E\, e^{i\omega t} \quad [4.82]$$

$$n = n_0 + \Delta n\, e^{i\omega t} \quad [4.83]$$

$$j = j_0 + \Delta I\, e^{i\omega t} \quad [4.84]$$

$$\Delta j_n(\vec{r},t) \;=\; e.N_D\,\mu\Delta E) + \varepsilon\mu\;\; E_0\frac{\partial \Delta E}{\partial x} + \varepsilon\mu\;\; \Delta E\;\; \frac{\partial E_0}{\partial x} + i\varepsilon\omega\Delta E \quad [4.85]$$

Hence:

$$\varepsilon\mu\;\; E_0\frac{\partial \Delta E}{\partial x} + (\,e\,\mu\;\; N_D + \varepsilon\mu\frac{\partial E_0}{\partial x} + i\varepsilon\omega)\Delta E = \Delta j_n \quad [4.86]$$

Si $n{=}N_D$

$$\varepsilon\mu E_0\frac{\partial \Delta E}{\partial x} + (-e\,\mu_n + i\varepsilon\omega)\Delta E = \Delta j_n \quad [4.87]$$

$$\varepsilon\mu E_0\frac{\partial \Delta E(x,x')}{\partial x} + (\,e\,\mu\;\; N_D + i\varepsilon\omega)\Delta E'x,x') = \Delta j_n \quad [4.88]$$

The solution is:

$$\Delta E(x,x') = \frac{\Delta j_n}{-e\,\mu_n + i\varepsilon\omega} + K(x)e^{-\frac{e\,\mu_n N_D + i\varepsilon\omega}{\epsilon\mu_0 E_0}x'} \quad [4.89]$$

$$\Delta E(x,x') = \frac{\Delta j_n}{e\,\mu_n + i\varepsilon\mu\omega} + (1 - e^{\frac{e\,\mu N_D + i\varepsilon\mu\omega}{\epsilon\mu_0 E_0}}(x - x')) \quad [4.90]$$

The impedance gradient is defined by:

$$\frac{\partial Z(N,x)}{\partial x} = -\frac{1}{\Delta I}\int_x^L \frac{\partial \Delta E(x,x")}{\partial x} dx' \quad [4.91]$$

Therefore, if the diffusion length is much smaller than that of the sample, we can write:

$$\frac{\partial Z(N,x)}{\partial x} = \frac{\Delta j_n}{S(\,i\varepsilon\omega^\circ\, e\,\mu_n)}.\left(1 - e^{\frac{e\,\mu\;\; N_D + i\varepsilon\omega}{\epsilon\mu_0 E_0}(x-L)}\right) \quad [4.92]$$

As for the noise,

$$\frac{\partial Z(N,x)}{\partial x} = \frac{\Delta j_n}{S(\,i\varepsilon\omega \mp e\,\mu_n D)} \quad [4.93]$$

Using the Einstein relationship, kT/q = D/μ [4.94]

We get:

$$S_v = 4e^2 D\mu.\frac{nL}{(S\varepsilon\omega)^2 + (Se\,\mu N_D)^2} \quad [4.95]$$

Finally,

$$S_v = 4kT\frac{R}{1+(\omega RC)^2} \quad [4.96]$$

It is the noise of the parallel RC circuit (see Chapter 3 – first-order electrical calculations for buried layers modeling the substrate, and second-order electrical calculations below), where

C=εS/L [4.97a]

and

R= L/ (qμnS) [4.97b]

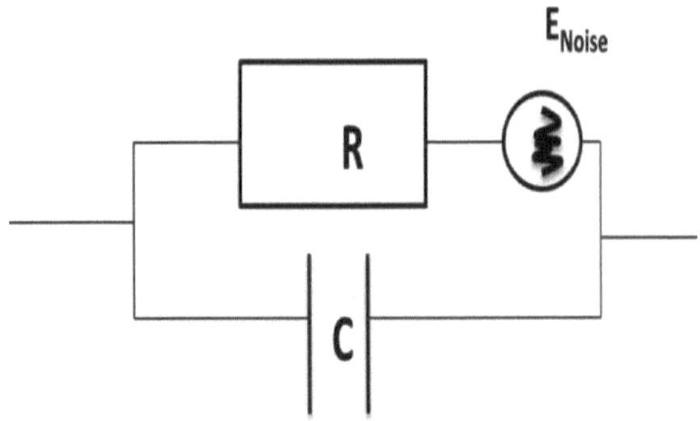

Figure 4.1. *Equivalent circuit with its noise source*

$$Zeq= (1/R+jC\omega)^{-1}= R/(1+jRC\omega)^{-1} \quad [4.98]$$

Thus:

$$S_V = 4kRT\frac{1}{1+(\omega RC)^2} \quad [4.99]$$

4.5. Noise calculation using the transfer impedance method principle: formulation [GHI 93, ROH 71, NOU 91]

We have observed in the impedance field method that a current generator in a dx slice induces a voltage variation at the junctions of the whole device.

As we have suggested, Green's algorithm may be linked to a transfer impedance, considering signal fluctuations. Injecting a current variation to a contact creates a potential variation at any other contract. Similarly, we must be able to analyze possible correlations between different noise sources; although these sources are autonomous, if they are very close – possible case in nanoelectronics – they can be correlated.

We consider another paradigm: the transfer impedance method (TIM); between two points, two slices or two volume elements, via a linear relationship between the electric field response and the potential difference at a current disturbance situated far away. Hence, we get the possibility of calculating internal noise spectra that come from these fluctuations. This method, generalizing the original impedance field method of Shockley *et al.*, was widely used in recent decades to calculate noise and is only efficient in one dimension. The important role of this method was, and still is, to calculate the entire range of voltage fluctuations between the junctions of the device. Currently, the problem remains unsolved in 3D (even in 2D), not because of numerical difficulties, but from a physical point of view.

4.5.1. *Principle*

A current generator induces a voltage variation in every dx slice. The result is that two distinct slices can be correlated. Thus, we cannot write that the spectral density (d.s.) of voltage at junctions of the whole set is the sum of d.s. of the voltage of each separate slice (see Salami method).

The $\Delta v(x)$ $\Delta v(x')$ cross terms must be added. More specifically, let us consider two slices, numbered as α and ß. Microscopic noise sources, each represented by a noise current generator denoted as Si_α and $Si_ß$, are untiled, but not the voltage fluctuations induced by these noise sources. The voltage induced at the junctions of slice ß by the current i_α is $Z_{\beta\alpha}(\omega).i_\alpha$, where $Z_{\beta\alpha}$ is the transfer impedance between slices α and ß. As a result, the total voltage induced at the junctions of ß for all the current generators of different slices is:

$$\Delta V_\beta = \sum_\alpha z_{\beta\alpha} i_\alpha \quad [4.100]$$

with i_α and i_β being untiled, we get:

$$S_v \Delta f = \sum_{\alpha\gamma}^{\infty} \overline{|i_\alpha{}^2|} z_{\beta\alpha} z_{\alpha\gamma}^* \quad [4.101]$$

Let us have

$$\vec{E}(\vec{r}) = \vec{E_0}(\vec{r}) + \delta\vec{E}(\vec{r},\omega)e^{i\omega t} \quad [4.102]$$

$$\vec{j}(\vec{r}) = \vec{j_0}(\vec{r}) + \delta\vec{j}(\vec{r},\omega)e^{i\omega t} \quad [4.103]$$

The first-order terms are linked via an operator $\hat{L}$:

$$\hat{L}\,\delta\vec{E}(\vec{r},\omega) = \delta\vec{j}(\vec{r},\omega) \quad [4.104]$$

We can then introduce Green's functions of the operator $\hat{L}$

$$\hat{L}\,\vec{\vec{z}}\left(\vec{r},\vec{r'},\omega\right) = \vec{\vec{1}}.\delta\left(\vec{r}-\vec{r'}\right) \quad [4.105]$$

where $\vec{\vec{1}}$ represents the unit matrix:

$$\delta\vec{E}(\vec{r},\omega) = \iiint \vec{\vec{z}}\left(\vec{r},\vec{r'},\omega\right)\delta\vec{j}(\vec{r'},\omega)d^3r' \quad [4.106]$$

The ddp at the junctions of a "dr" on an interelectrode line can be written as:

$$\delta V^2 = -\int_0^L dr^T \iiint \vec{\vec{z}}(\vec{r}, \vec{r'}, \omega)\, \delta\vec{j}(\vec{r'}, \omega) d^3r' \qquad [4.107]$$

Therefore, between two electrodes:

$$\delta V = -\int_0^L dr^T \iiint \vec{\vec{z}}(\vec{r}, \vec{r'}, \omega)\, \delta\vec{j}(\vec{r'}, \omega) d^3r' \qquad [4.108]$$

Finally, for a material of length L, after some simple calculation that may be a little "heavy", we get:

$$S_V \Delta f(\omega) = \int_0^L \int_0^L \iiint d^3r'' \iiint d^3r''' dr^T \vec{\vec{z}}(\vec{r}, \vec{r''}, \omega)\, \vec{j_N}(\vec{r''}, \omega).\vec{j_N}'(\vec{r'''}, \omega) \vec{\vec{z^T}}(\vec{r'}, \vec{r'''}, \omega)\, dr' \qquad [4.109]$$

We also have the spectral density of current fluctuation:

$$\vec{j_N}(\vec{r''}, \omega).\vec{j_N}'(\vec{r'''}, \omega) = K \quad (\vec{r''}, \omega).\,\delta\left(\vec{r''} - \vec{r'''}\right)\Delta f \qquad [4.110]$$

with K being the noise source.

If we want to get "ddps" or impedance of the whole sample, curvilinear integrals must be calculated. Currently, some calculations can be carried out only in 1D, which is a big hurdle.

Equation [4.109] is written as:

$$S_V(\omega) = \iint \vec{\nabla} Z \quad (r, \omega) \cdot Sj(, r \quad , \omega). \quad \nabla Z^T dr^3 \qquad [4.111]$$

For a comparison with [4.75], see the impedance field.

NOTE.– $S_V(\omega)$ is zero if the Z gradient is zero. Hence, we can create a (straight) strongly and uniformly doped slice between the noise source and a target to be protected.

4.5.2. *One-dimensional structures*

4.5.2.1. *Noise of a bar in the ohmic regime at high frequencies*

$$E = E_0 + \delta E\, e^{j\omega t} \qquad [4.112]$$

$$n = n_0 + \delta n\, e^{j\omega t} \quad [4.113]$$

$$I = I_0 + \delta I\, e^{j\omega t} \quad [4.114]$$

$$j = q\mu n E + \varepsilon \frac{\partial E}{\partial t} \quad [4.115]$$

$$\frac{\partial E}{\partial x} = \frac{-q}{\varepsilon}(n - N_d) \quad [4.116]$$

At high frequencies, we can write:

$$\frac{\delta v}{\delta E} = \frac{\mu_0}{1 + i\omega\tau} \quad [4.117]$$

NOTE.– At very low frequencies, $\frac{\delta v}{\delta E}$ tends towards μ_0 and in the ohmic regime, $\frac{\partial E}{\partial x}$ =0; thus, n=Nd

$$\delta I = -qA(n_0\, \delta v + v_0\, \delta n) + ieA\omega\, \delta E \quad [4.118]$$

where A is a square section

$$\delta I = \left(\frac{qAn_0\mu_0}{1+i\omega\tau} + i\omega A\right)\delta E - qA\, v_0\, \delta n) \quad [4.119]$$

$$\delta I = \left(\frac{qAn_0\mu_0}{1+i\omega\tau} + i\omega A\right)\delta E - \epsilon A\mu_0 E_0 \frac{\partial \delta E}{\partial x} \quad [4.120]$$

For transfer impedance:

$$-\epsilon A\mu_0 E \frac{\partial z(x,x,\omega)}{\partial x} + \left(\frac{qAn_0\mu_0}{1+i\omega\tau} + i\omega\epsilon A\right)z(x,x,\omega) = \delta' x - x') \quad [4.121]$$

The solution for this differential equation is then:

$$z(x,x',\omega) = Ke^{\frac{qn_0\mu_0 - \omega^2\tau\epsilon + i\omega\epsilon_0}{\epsilon\mu_0 E_0}}x \quad [4.122]$$

For example, through the variation of constants theorem, we get:

$$K = -\frac{1 + i\omega\tau}{AE_0\epsilon\mu_0} + i\omega\epsilon A e^{-\lambda x'} H(x-x') + C \quad [4.123a]$$

where H is the Heaviside step function.

In addition:

$$\lambda = \frac{qn_0\mu_0 - \omega^2\,\tau\epsilon + i\omega\epsilon}{\epsilon_0\mu_0 E_0} \qquad [4.123b]$$

Introducing δI in x' results in x>x' if we ignore the diffusion.

In short:

$$z(x, x', \omega) = -\frac{1+i\omega\tau}{A\epsilon\ \mu_0 E_0} e^{\frac{(qn_0\mu_0 - \omega^2\,\tau\epsilon + i\omega\epsilon)(x-x')}{\epsilon\mu_0 E_0}} H(x - x') \qquad [4.124]$$

Hence, we have its gradient:

$$\vec{\nabla} z(x', \omega) = \frac{1+i\omega\tau}{A\epsilon_0\mu_0 E_0}\left(1 - e^{\frac{(qn_0\mu_0 - \omega^2\,\tau\epsilon + i\omega\epsilon).(L-x')}{\epsilon\mu_0 E_0}}\right) \qquad [4.125]$$

In the ohmic regime (E constant, very weak): $\lambda \to \infty$, we get:

$$lim_{E_0 \dashrightarrow 0} \vec{\nabla} z(x', \omega) = \frac{1+i\omega\tau}{A(\,qn_0\mu_0 - \omega^2\,\tau\epsilon + i\omega\epsilon)} \qquad [4.126]$$

The diffusion noise power is thus:

$$S_V = 4q^2 A \int_0^L n_0 D_0 |\nabla Z|^2\, dx \qquad [4.127]$$

Yet, in the ohmic regime, $kT/q = D_0/\mu_0$

Thus,

$$S_V = 4\frac{kT}{Aqn_0\mu_0} L = 4kT\frac{L}{\sigma_0 A} = 4kRT$$

with σ_0 being the ohmic conductivity.

We obtain the well-known Nyquist–Johnson relationship.

NOTE.– The approach for calculating the impedance field, in practice, can be to force the AC sources at each mesh node of the device and evaluate the response at the device's electrodes.

A method derived from the system principle is simpler to implement; we calculate the transimpedance Z_{ij}, where i is any node and j is an electrode. As a result, $Z_{ij} = Z_{ji}$. It can be determined by applying an AC source at a terminal and calculating the internal potentials at each node [GHI 93, ROH 71, LEE 96, BON 98].

4.6. Substrate noise: towards 3D

The generation–recombination noise (GR) represents a noise source that is typical and fundamental of a noise in semiconductor materials and related devices, where the carrier concentration can considerably change the order of magnitude. For electron density, fluctuations can occur between two levels, namely conduction band and donor impurity, or due to defects. Conductivity, associated with GR noise, can be written by relating it to an average carrier lifespan τ:

$$\sigma_{GR} = \frac{\sigma_0}{1+j\omega\tau} \qquad [4.128]$$

Let us consider a typical 0.35 µm BiCMOS technology (see Figure 3.66, Chapter 3), a particular region (P+/P) is shown. The real profile is modeled by a stack of layers of uniform thickness (taken *ex abrupto*; see Chapter 3) and uniformly doped.

Let us consider two places where we see this GR:

a) "bad" contact

contact C4 (see Figure 3.37), at the base, in the third layer L_3: GR, or

b) GR in the whole of layer L_6

Modulations of impedance resulting in these GR noises are shown in Figure 4.2(a) and 4.2(b). Our first results are qualitative, and also seem quantitatively realistic. For instance, we see, in both cases, the influence of the additional GR current, reducing the resistance to low frequencies, before a cut-off frequency of f_c=1/(2πτ). The impedance for a lifespan of τ of 1 ns can be reduced by 5% of its reference value, without GR.

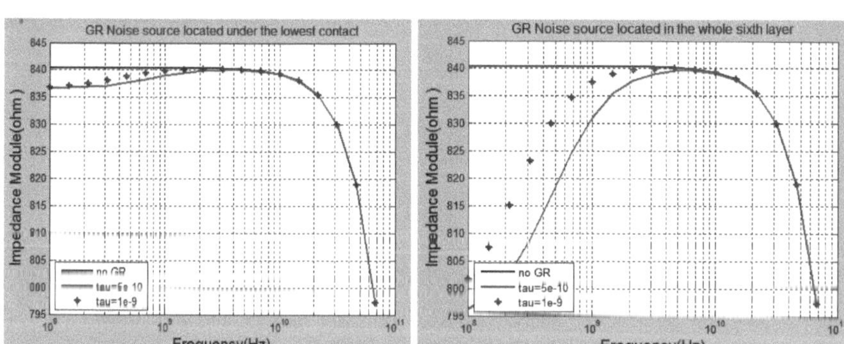

Figure 4.2. *Introduction of a GR process: (a) defaults under contact C4 and (b) defaults in the sixth layer (see Figure 3.66, Chapter 3). For a color version of this figure, see www.iste.co.uk/gontrand/nanoelectronics.zip*

Finally, Figure 4.3 presents the noise spectral densities of previous two GR cases. For this, we use the impedance transfer paradigm, applied to the multilayer substrate (see Figure 4.3), using equations such as [4.111].

It should be possible to study current noise, induced by temperature fluctuations in the volume, using these methods.

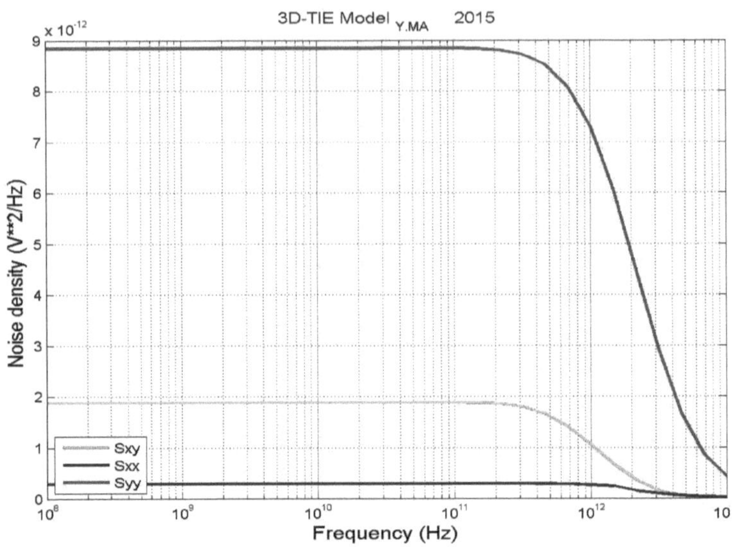

Figure 4.3. *Noise spectral densities (see Figure 3.66, Chapter 3). For a color version of this figure, see www.iste.co.uk/gontrand/nanoelectronics.zip*

4.7. Appendix

The DDM must be applied to the fluctuation of physical and electrical variables in components.

4.7.1. *Calculation:* δn, δp, $\delta \Phi$

– First, the DDM software must be applied to determine the quiescent point that we will denote as (Φ_0, P_0, n_0).

– Second, a modulus, which allows for calculating variations (δn, δp, $\delta \Phi$) that are induced by the presence of defaults, must be developed.

4.7.1.1. *First solution*

4.7.1.1.1. Poisson's equation

Quiescent point (Φ_0, P_0, n_0) [4.129]

$$\Delta(\Phi_0) = \frac{q}{\varepsilon_{c/s}}(n_0 - p_0 - dop) \quad [4.130]$$

N_T is zero at the quiescent point.

where:

– n and p are carrier densities;

– q is the electrical load;

– $\zeta_{s/c}$ is the prematurity of the semiconductor;

– Φ is the electrostatic potential.

4.7.1.2. *Changes made to equations*

$$\Delta(\Phi_0 + \delta\Phi) = (n_0 + \delta n - p_0 - \delta p + dop + N_T) \quad [4.131]$$

$$\begin{aligned}\Delta(\Phi_0 + \delta\Phi) - \Delta(\Phi_0) = (&N_0 + \delta n - p_0 - \delta p + dop \\ &+ N_T - N_0 + P_0 - dop)\end{aligned} \quad [4.132]$$

$$\Delta(\delta\Phi) = (\delta n - \delta p + N_T) \quad [4.133]$$

We introduce a value of N_T smaller than 10^{18} cm^{-3} of the standard doping of a base, or of order 10^{15} cm^{-3} if the N_T are at the base level.

4.7.1.3. *Φ_n and Φ_p pseudo-Fermi levels*

We have a set of three equations within three unknown values (Φ_n, Φ_p, Φ), (n, p, Φ).

If we now consider variations for the above variables, leading to the following triplets:

$$(\Phi_n + \delta\Phi_n,\ \Phi_p + \delta\Phi_p,\ \Phi + \delta\Phi)\ (n_0 + \delta n,\ p_0 + \delta p,\ \Phi_0 + \delta\Phi). \quad [4.134]$$

$$\Delta(\phi_n) - 0 \quad \Delta(\phi_p) = 0 \quad [4.135]$$

The variation is $\delta\Phi_n$ and $\delta\Phi_p$:

$$\Delta\left(\phi_n + \delta\phi_n\right) = 0 \text{ and } \Delta\left(\phi_p + \delta\phi_p\right) = 0 \qquad [4.136]$$

At the boundary conditions, we have

Φ_a being the value applied at contacts

$\phi_n = \phi_a$ at contacts: collector, base, emitter

$\phi_n + \delta\phi_n = \phi_a$, which means $\delta\phi_n = 0$

$$\begin{cases} E_{C0} + \delta E_c = -\Phi_0 - \delta\Phi_0 + \dfrac{E_g}{2} & [4.137a] \\ E_{v0} + \delta E_v = -\Phi_0 - \delta\Phi_0 - \dfrac{E_g}{2} & [4.137b] \end{cases}$$

This means that:

$$\delta E_c = -\delta\Phi_0 \qquad [4.138a]$$

$$\delta E_v = \delta\Phi_0 \qquad [4.138b]$$

Carrier densities can then be written as:

$$n_0 = N_c \exp(-\Phi_{n0}) \exp\left(-E_{C0}\right) \qquad [4.139a]$$

$$p_0 = N_v \exp(\Phi_{p0}) \exp\left(E_{v0}\right) \qquad [4.139b]$$

For a variation δn and δp, n and p are as follows:

$$\begin{cases} n_0 + \delta n = N_c \exp(-\Phi_{n0} - \delta\Phi_n) \exp\left(-E_{C0} - \delta E_c\right) & [4.140a] \\ p_0 + \delta p = N_v \exp(\Phi_{p0} + \delta\Phi_p) \exp\left(E_{v0} + \delta E_v\right) & [4.140b] \end{cases}$$

$$\begin{cases} n_0 + \delta n - n_0 = N_c \exp(-\Phi_{n0} - \delta\Phi_n)\exp(-E_{C0} - \delta E_c) \\ \qquad -N_c \exp(-\Phi_{n0})\exp(-E_{c0}) & [4.141a] \\ p_0 + \delta p - p_0 = N_v \exp(\Phi_{p0} + \Phi_n)\exp(E_{v0} + \delta E_v) \\ \qquad -N_v \exp(\Phi_{p0})\exp(E_{v0}) & [4.141b] \end{cases}$$

$$\begin{cases} \delta n - N_c \exp(-\Phi_{n0})\exp(-E_{c0}) \left[(\exp(-\delta\Phi_n)\exp(-\delta E_c)) - 1\right] & [4.142a] \\ \delta p = N_v \exp(\Phi_{p0})\exp(+E_{v0}) \left[((\delta\Phi_p)\exp(+\delta E_v)) - 1\right] & [4.142b] \end{cases}$$

$$\begin{cases} \delta n = N_c \exp(-\Phi_{n0})\exp(-E_{c0}) \left[(\exp(-\delta\Phi_n - \delta E_c)) - 1\right] & [4.143a] \\ \delta p = N_v \exp(\Phi p_0)\exp(E_{V0}) \left[(\exp(\delta\Phi_n + \delta E_c)) - 1\right] & [4.143b] \end{cases}$$

Variations δn and δp are due to the variation $\delta\phi$ (and vice versa).

At contacts:

For potential $\delta\phi$:

$$\Phi_0 + \delta\Phi = phic + (kT/q).\ln(n_0 + \delta n)$$

$$\Phi_0 + \delta\Phi - \Phi_0 = phic + \ln(n_0 + \delta n) - phic - \ln(n_0)$$

$$\delta\Phi = (kT/q).\ln(n_0 + \delta n) - (kT/q).\ln(n_0) \sim (kT/q).\ dn/n_0 \qquad [4.144]$$

– The variation of δn and δp at the contacts is zero:

$$n_0 = \frac{dop + \sqrt{4 + dop}}{2} \text{ normalized by } n_i$$

$$p_0 = \frac{1}{n_0} \text{ normalized by } n_i$$

– It means that $\delta n = 0$, so $\delta\phi$ is zero.

At base–emitter–collector contacts:

$$\delta\Phi = 0 \quad \delta n = 0 \quad \delta p = 0 \qquad [4.145]$$

4.7.1.4. *Final solution*

We will now solve Poisson and continuity equations for electrons and holes.

4.7.1.4.1. Poisson

$$\Delta(\delta\Phi) = (\delta n - \delta p + N_T) \qquad [4.146]$$

We have the approximate δn and δp from the first solution.

4.7.1.4.2. Continuity equations for electrons and holes

$$\begin{cases} \vec{J}_{n0} = q.\mu_n.n_0.\vec{E}_0 + q.D_n.\vec{\nabla} n_0 & [4.147a] \\ \vec{J}_{p0} = q.\mu_p.p_0.\vec{E}_0 - q.D_p.\vec{\nabla} P_0 & [4.147b] \end{cases}$$

where:

– Jn_0 is the electron current density;

– J_{p0} is the hole current density;

– μ_n, μ_p are electron and hole mobilities in a semiconductor;

– D_n, D_p are diffusion coefficients in a semiconductor or material;

– E_{01} is the electric field.

$$\begin{cases} \dfrac{\partial n_0}{\partial t} = GR_n + \dfrac{1}{q}\vec{\nabla} J_{n0} & [4.148a] \\ \dfrac{\partial p_0}{\partial t} = GR_p - \dfrac{1}{q}\vec{\nabla} J_p & [4.148b] \end{cases}$$

$$\begin{cases} \text{In static case } \dfrac{\partial n}{\partial t} = 0 \end{cases} \qquad [4.149]$$

The generation–recombination term is written as G_{SHR} (Shockley, Hall, Read):

$$GR_{n0} = GR_{p0} = G_{SHR} = \frac{n_0 \cdot p_0 - n_i^2}{\tau_n\left(p_0 + p_i\right) + \tau_p\left(n_0 + n_i\right)} \qquad [4.150]$$

$$\begin{cases} -\frac{1}{q}\vec{\nabla}\left(J_{n0}+\delta J_n\right)=G_{SHR0}+G_{SHRdef} & \text{[4.151a]} \\ \frac{1}{q}\vec{\nabla}\left(J_{p_0}+\delta J_p\right)=G_{SHR0}+G_{SHRdef} & \text{[4.151b]} \end{cases}$$

$$\begin{cases} \vec{J}_{n0}+\delta\vec{J}_n=q.\mu_n.n_0\vec{E}_0+q\mu_n\delta n\,\vec{E}_0+q\mu_n n_0\,\delta\vec{E} \\ \qquad +q\mu_n\,\delta n\,\delta\vec{E}+q.D_n.\vec{\nabla}(n_0+\delta n) & \text{[4.152a]} \\ \vec{J}_{P0}+\delta\vec{J}_p=q.\mu_p.p_0\vec{E}_0+q\mu_p\delta p\quad\vec{E}_0+q\mu_p p_0\delta\vec{E} \\ \qquad +q\mu_p\delta p\delta\vec{E}-q.D_p.\vec{\nabla}(p_0+\delta p) & \text{[4.152b]} \end{cases}$$

$$\begin{cases} \vec{J}_{n0}+\delta\vec{J}_n=q.\mu_n.n_0\vec{E}_0+q\mu_n\delta n\,\vec{E}_0+q\mu_n n_0\,\delta\vec{E} \\ \qquad +q\mu_n\,\delta n\,\delta\vec{E}+q.D_n.\vec{\nabla}n_0+qD_n\vec{\nabla}\,\delta n & \text{[4.153a]} \\ \vec{J}_{P0}+\delta\vec{J}_p=q.\mu_p.p_0\vec{E}_0+q\mu_p\delta p\quad\vec{E}_0+q\mu_p p_0\delta\vec{E} \\ \qquad +q\mu_p\delta p\delta\vec{E}-q.D_p.\vec{\nabla}p_0-D_p\vec{\nabla}\delta p & \text{[4.153b]} \end{cases}$$

We ignored the terms of the second-order (δ. δ) linked to the following expressions: $q\mu_n\,\delta n\,\delta\vec{E}$ and $q\mu_p\,\delta p\,\delta\vec{E}$ and

$$\begin{cases} \vec{J}_{n0}+\delta\vec{J}_n-\vec{J}_{n0}=\delta\vec{J}_n=q\mu_n\delta n\,\vec{E}_0+q\mu_n n_0\,\delta\vec{E} \\ \qquad +q\mu_n\,\delta n\,\delta\vec{E}+qD_n\vec{\nabla}\,\delta n & \text{[4.154a]} \\ \vec{J}_{P0}+\delta\vec{J}_p-\vec{J}_{p0}=\delta\overrightarrow{J_p}=q\mu_p\delta p\quad\vec{E}_0+q\mu_p p_0\delta\vec{E} \\ \qquad +q\mu_p\delta p\delta\vec{E}-D_p\vec{\nabla}\delta p & \text{[4.154b]} \end{cases}$$

$$G_{SHRdef}=\frac{(n_0+\delta n).\ (p_0+\delta p)-n_i^2}{\tau_P\left(n_0+\delta n+n_i\exp\left(\frac{E_T-E_F}{KT}\right)\right)+\tau_n\left(p_0+\delta p+n_i\exp\left(\frac{E_F-E_T}{KT}\right)\right)} \quad \text{[4.155a]}$$

$$G_{SHRdef} = \frac{n_0\, p_0 + \delta n\, p_0 + \delta p\, n_0 + \delta n\, \delta p - n_i^2}{\tau_P\left(n + \delta n + n_i \exp\left(\frac{E_T - E_F}{KT}\right)\right) + \tau_n\left(p + \delta p + n_i \exp\left(\frac{E_F - E_T}{KT}\right)\right)} \quad [4.155b]$$

$$\begin{cases} \delta \vec{J}_n = q\mu_n \delta n\, \vec{E}_0 + q\mu_n n_0\, \delta\vec{E} + q\mu_n\, \delta n\, \delta\vec{E} + qD_n \vec{\nabla}\, \delta n & [4.156a] \\ \delta\overrightarrow{J_p} = q\mu_p \delta p\;\; \vec{E}_0 + q\mu_p p_0 \delta\vec{E} + q\mu_p \delta p \delta\vec{E} - D_p \vec{\nabla}\delta p & [4.156b] \end{cases}$$

From the first solution, we have: δp and δn , $\delta\phi$. Hence, we can find δE

$$\begin{cases} -\frac{1}{q}\vec{\nabla}\cdot\left(\vec{J}_{n0} + \delta\overrightarrow{J_n}\right) + \frac{1}{q}\overrightarrow{\nabla}\cdot\left(\overrightarrow{J_{n0}}\right) = G_{SHR0} + G_{SHRdef} - G_{SHR0} & [4.157a] \\ \frac{1}{q}\vec{\nabla}\left(\overrightarrow{J_{p0}} + \delta\overrightarrow{J_p}\right) - \frac{1}{q}\vec{\nabla}\left(\overrightarrow{J_{p0}}\right) = G_{SHR0} + G_{SHRdef} - G_{SHR0} & [4.157b] \end{cases}$$

$$\begin{cases} -\frac{1}{q}\vec{\nabla}\,\delta J_n = G_{SHRdef} & [4.158a] \\ \frac{1}{q}\vec{\nabla}\,\delta J_p = G_{SHRdef} & [4.158b] \end{cases}$$

Substituting equations [4.156a] and [4.156b] into [4.158a] and [4.158b], respectively:

$$\begin{cases} -\frac{1}{q}\vec{\nabla}\left(q\mu_n \delta n \vec{E}_0 + q\mu_n n_0\, \delta\vec{E} + q\mu_n\, \delta n \delta\vec{E} + qD_n \vec{\nabla}\, \delta n\right) = G_{SHRdef} & [4.159a] \\ \frac{1}{q}\vec{\nabla}\left(q\mu_p \delta p\;\; \vec{E}_0 + q\mu_p p_0 \delta\vec{E} + q\mu_p \delta p \delta\vec{E} - D_p \vec{\nabla}\delta p\right) = G_{SHRdef} & [4.159b] \end{cases}$$

Therefore, we will have a system of three equations coupled with three unknown values (δp , δn , $\delta\phi$) (Gummel algorithm):

$$\Delta(\delta\Phi) = (\delta n - \delta p + N_T) \qquad [4.160]$$

$$\begin{cases} -\dfrac{1}{q}\vec{\nabla}\left(q\mu_n\delta n\vec{E}_0 + q\mu_n n_0\,\delta\vec{E} + q\mu_n\,\delta n\,\delta\vec{E} + qD_n\vec{\nabla}\delta n\right) = G_{SHRdef} & [4.161] \\ \dfrac{1}{q}\vec{\nabla}\left(q\mu_p\delta p\;\;\vec{E}_0 + q\mu_p p_0\delta\vec{E} + q\mu_p\delta p\delta\vec{E} - D_p\vec{\nabla}\delta p\right) = G_{SHRdef} & [4.162] \end{cases}$$

4.7.1.5. *Observations*

– $\delta\vec{J}_n = q\mu_n\delta n\,\vec{E}_0 + q\mu_n n_0\,\delta\vec{E} + q\mu_n\,\delta n\,\delta\vec{E} + qD_n\vec{\nabla}\,\delta n$

– $\delta\overrightarrow{J_p} = q\mu_p\delta p\;\;\vec{E}_0 + q\mu_p p_0\delta\vec{E} + q\mu_p\delta p\delta\vec{E} - D_p\vec{\nabla}\delta p$

Even though δp, δn are initially zero, $\delta J_n \neq 0$ and $\delta J_p \neq 0$ because $\delta E \neq 0$:

$$\delta E = \frac{\delta(\Phi_i - \Phi_{i+1})}{\Delta x} = \frac{\delta\Phi_i}{\Delta x} - \frac{\delta\Phi_{i+1}}{\Delta x} \qquad [4.163]$$

4.8. Conclusion

Modeling in 3D noise (even 2D) is mostly a failure due to physical modeling problems: for example, discerning the hole and electron contribution to noise in devices. Master equations, like the Boltzmann equation, associated with second-order electrical studies, should be developed more for a theoretical study of noise.

For submicronic devices, new effects, like non-stationary phenomena, crop up. By reducing the length of a device to some mean free paths, each carrier undergoes very few collisions on its path from one electrode to another. Another example is the correlation of noise sources in space, which are not independent, even though they are autonomous. Moreover, intensities of noise can increase during voltage ramp applied to a circuit. Other investigations must be made, in the case of the following points:

– quantum phenomena: duration of carrier collisions versus energy exchange with phonons;

– proposing rigorous techniques or methods of noise reduction, although it is subtler than it seems in this ambiguous domain.

The future program to be developed could focus on the following main areas:

– continuing basic studies on the microscopic behavior of noise sources up to 3D, with theories and experiments, in order to have a better understanding of phenomena;

– study on submicronic, nanometric transistors, used for high-speed logic in information technology. The objective is to study the 1/f noise and RTS (Random Telegraphic Signal) noise, in order to better understand ultimate (Bi-)CMOS and develop compact electric models of component noise to correctly and fully simulate the behavior of these devices and related circuits;

– high-speed III–V transistors (HEMT and HBT), as important elements of future high-frequency broadband telecommunication systems. Low-frequency analyses (1/f, GR and RTS noise) and high-frequency noise will be conducted on semiconductor devices and material (bulk, epitaxial layers, flexible electronics);

– noise in photoelectric detectors and lasers, particularly in multi-quantum well structures, for high-speed, optoelectronics and gigabit communications;

– sensors, in particular those using magnetic effects.

Bibliography

[ADD 66] ADDA Y., PHILIBERT J., *La diffusion dans les solides*, vols 1 and 2, Bibliothèque des Sciences et Techniques Nucléaires, Presses Universitaires de France, Paris, 1966.

[BAR 99] BARBIERI S., BELTRAM F., ROSSI F., "Microscopic theory of vertical-transport phenomena in semiconductor heterostructures: interplay between two- and three-dimensional hot-carrier relaxation", *Physical Review B*, vol. 60, p. 1953, 1999.

[BIE 75] BIERSACK J.P., FINK D., in NAMBA S. (ed.), *Ion Implantation in Semiconductors*, Plenum Press, New York, pp. 211–218, 1975.

[BOH 51] BOHM D., *Quantum Theory*, Prentice-Hall, New York, 1951.

[BON 98] BONANI F., GHIONE G., PINTO M.R. *et al.*, "An efficient approach to noise analysis through multidimensional physics-based models", *IEEE Transactions on Electron Devices*, vol. 45, pp. 261–269, 1998.

[BOR 99] BORDONE P., PASCOLI M., BRUNETTI R. *et al.*, "Quantum transport of electrons in open nanostructures with the Wigner-function formalism", *Physical Review B*, vol. 59, no. 4, pp. 3060–3069, 1999.

[BOU 03] BOUCARD F., Modélisation de la diffusion des dopants dans le silicium pour la réalisation de jonctions fines, PhD Thesis, Louis Pasteur University, Strasbourg, 2003.

[BOU 13] BOUNOUAR M.A., Double-gate single electron transistor: modeling, design & evaluation of logic architectures, PhD Thesis, University of Sherbrooke, Quebec, Canada and Institut des Nanotechnologies de Lyon (INL), INSA de Lyon, France, 2013.

[CER 88] CERCIGNANI C., *The Boltzmann Equation and its Applications*, 2nd ed., Springer, New York, 1988.

[CHA 74] CHANG L.L., ESAKI L., TSU R., "Resonant tunneling in semiconductor double barriers", *Applied Physics Letters*, vol. 24, p. 593, 1974.

[CHO 85] CHO K., NUMAN M., FINSDTAD T.G. *et al.*, "Transient enhanced diffusion during rapid thermal annealing of boron implanted silicon", *Applied Physics Letters*, vol. 47, p. 1723, 1985.

[CIA 96] CIAMPOLINI P., ROSELLI L., STOPPONI G., "Mixed-mode circuit simulation with full-wave analysis of interconnections", *Proceedings of the 26th European Solid State Device Research Conference, ESSDERC '96*, IEEE Conference Publications, pp. 487–490, 1996.

[CLE 01] CLERC R., SULLIVAN P.O., MC CARTHY K.G. *et al.*, "A physical compact model for direct tunneling from NMOS inversion layers", *Solid-State Electronics*, vol. 45, pp. 1705–1716, 2001.

[COM] COMSOL, http://www.comsol.com.

[CRA 75] CRANCK J., *The Mathematics of Diffusion*, 2nd ed., Clarendon Press, Oxford, 1975.

[CRE 98] CRESSLER J.D., "SiGe TBH technology: a new contender for Si-based RF and microwave circuit applications", *IEEE Transactions on Microwave Theory and Techniques*, vol. 46, no. 5, pp. 572–589, 1998.

[DAT 00] DATTA S., "Nanoscale device modeling: the Green's function method", *Superlattices and Microstructures*, vol. 28, no. 4, pp. 253–278, 2000.

[FAI 84] FAIR R.B., WORTMAN J.J., LIU J., "Modeling rapid thermal diffusion of arsenic and boron in silicon", *Journal of the Electrochemical Society*, vol. 131, p. 2387, 1984.

[FAI 86] FAIR R.B., "Modeling of dopant diffusion during rapid thermal annealing", *Journal of Vacuum Science & Technology A: Vacuum, Surfaces, and Films*, vol. 4, p. 926, 1986.

[FIS 99] FISCHETTI M.V., "Master-equation approach to the study of electronic transport in small semiconductor devices", *Physical Review B*, vol. 59, no. 7, pp. 4901–4917, 1999.

[GED 96] GEDNEY S.D., "An anisotropic perfectly matched layer-absorbing medium for the truncation of FDTD lattices", *IEEE Transactions on Antennas Propagation*, vol. 44, no. 12, pp. 1630–1639, 1996.

[GHA 88] GHATAK A.K., THYAGARAJAN K., SHENOY M.R., "A novel numerical technique for solving the one-dimensional Schroedinger equation using matrix approach-application to quantum well structures", *IEEE Journal of Quantum Electronics*, vol. 24, no. 8, p. 1524, 1988.

[GHA 95] GHARPUREY R., MEYER R.G., "Modeling and analysis of substrate coupling in ICs", *Proceedings of the IEEE CICC*, pp. 125–128, May 1995.

[GHI 93] GHIONE G., FILICORI F., "A computationally efficient unified approach to the numerical analysis of the sensitivity and noise of semiconductor devices", *IEEE Transactions on Computer-Aided Design of Integrated Circuits and Systems*, vol. 12, no. 3, pp. 425–438, 1993.

[GON 95] GONTRAND C., HADDAB Y., KAABI L. *et al.*, "A new model of codiffusion in polycrystalline silicon specified to BiCMOS technology", *Semiconductor Science and Technology*, vol. 10, pp. 1393–1403, 1995.

[GON 06] GONZALEZ O., PEREDA J.A., HERRERA A. *et al.*, "An extension of the lumped-network FDTD method to linear two-port lumped circuits", *IEEE Transaction on Microwave Theory and Techniques*, vol. 54, pp. 3045–3051, 2006.

[HU 69] HU S.M., "General theory of impurity diffusion in semiconductors via the vacancy mechanism", *Physical Review*, vol. 189, pp. 773–784, 1969.

[ISE] ISE, SENTAURUS, "Tools for stimulating device performance", available at: https://www.synopsys.

[JIN 94] JINDAL R.P. (ed.)., "Special issue on fluctuation phenomena in electronic and, photonic device", *IEEE Transactions on Electron Devices*, vol. 41, pp. 2133–2138, 1994.

[KEY] KEYSIGHT TECHNOLOGIES, Advanced Design System (ADS), available at: http://www.keysight.com/en/pc-1297113/advanced-design-system-ads?cc=US&lc=eng.

[KUR 61] KURT L., ALEXIS S., "Diffusion of charged particles into a semiconductor under consideration of the built-in field", *Solid State Electronics*, vol. 3, no. 1, pp. 45–50, 1961.

[LAB 11] LABIOD S., LATRECHE S., BELLA M. *et al.*, "Combined electromagnetic and drift diffusion models for microwave semiconductor device", *Journal of Electromagnetic Analysis and Applications*, vol. 3, pp. 423–429, 2011.

[LAK 08] LAKHDARA M., LATRECHE S., GONTRAND C., "Modeling of parasitic effects induced by electrically active defects in a SiGe TBH", *European Physical Journal Applied Physics*, vol. 43, pp. 55–63, 2008.

[LAN 08] LANGEVIN P., "Sur la théorie du mouvement brownien", *Comptes Rendus de l'Académie des Sciences (Paris)*, vol. 146, pp. 530–533, 1908.

[LAN 75] LANFORD III O.E., "Time evolution of large classical systems", *Dynamical Systems, Theory and Applications*, Lecture Notes in Physics, vol. 38, Springer, Berlin, pp. 1–111, 1975.

[LAN 96] LANZEROTTI L.D., AMOUR A.S., LIU C. *et al.*, "$Si/Si_{1-x-y}Ge_xC_y/Si$ heterojunction bipolar transistor", *IEEE Electron Device Letters*, pp. 334–337, 1996.

[LEE 96] LEE J.B., MIN H.S., PARK Y.J., *Journal of Applied Physics*, vol. 79, no. 1, p. 228, 1996.

[LI 08] LI J., MIAO C., "A new implementation of the uniaxial perfectly matched layer", *Microwave and Millimeter Wave Technology*, vol. 2, pp. 770–773, 2008.

[LIN 63] LINDHARD J., SCHARFF M., SCHIOTT H.E., "Range concepts and heavy ion ranges", *Mat. Fys. Medd. Dan. Vid. Selsk*, vol. 33, no. 14, 42 pp., 1963.

[MA 16] MA Y., FAKRI-BOUCHET L.F., CALMON F. *et al.*, "Electrothermal modelling for three-dimensional nanoscale circuit substrates; noise", *IEEE Transactions on Components, Packaging and Manufacturing Technology*, vol. 6, no. 7, 2016.

[MAN 81] MANDURAH M.M., SARASWAT K.C., KAMINS T.I., "A model for conduction in polycrystalline silicon—part I: theory", *IEEE Transactions on Electron Devices*, vol. 28, no. 10, pp. 1163–1171, 1981.

[MAS 08] MASCART M., "Sur la théorie du mouvement brownien. Note de M. P. Langevin", *Comptes rendus hebdomadaires des séances de l'Académie des sciences, série physique*, session on 9 March 1908, vol. 146, pp. 530–533, 1908.

[MAT 91] MATHIOT D., "Hydrogen diffusion and passivation processes in p- and n-type crystalline silicon", *Physical Review B*, vol. 44, p. 6141, 1991.

[MAT 96] MATHIOT D., "Composants pour l'électronique: Dopage des semi-conducteurs", in PHILIBERT J., SABIONI A.C.S., DYMENT F. (eds), *Diffusion in Materials*, Editora REM-Revista Escola de Minas, Ouro Preto, 1996.

[MIC 87] MICHEL A.E., RAUSCH W., RONSHEIM P., "An implantation damage and the anomalous transient diffusion of ion-implanted boron", *Applied Physics Letters*, vol. 51, no. 7, pp. 416–418, 1987.

[MIR 10] MIRZAVAND R., ABDIPOUR A., MORADI G., "Full-wave semiconductor devices simulation using ADI-FDTD method", *Progress in Electromagnetic Research M*, pp. 191–202, 2010.

[MOG 90] MOGLESTUE C., "Self-consistent Monte Carlo particle modelling of small semiconductor elements", *Reports on Progress in Physics*, vol. 53, no. 10, pp. 1333–1339, 1990.

[MOO 65] MOORE G.E., "Cramming more components on to integrated circuits", *Electronics*, vol. 38, no. 8, p. 4, 1965.

[NIK 98] NIKNEJAD A.M., GHARPUREY R., MEYER R.G., "Numerically stable green function for modeling and analysis of substrate coupling in integrated circuits", *IEEE Transactions on Computer-Aided Design of Integrated Circuits and Systems*, vol. 17, no. 4, pp. 305–315, 1998.

[NOU 91] NOUGIER J.P., *IV-V Microelectronics*, Elsevier Science Publishers E.V., 1991.

[ORI 99] ORIOLS X., GARCIA-GARCIA J.J., MARTIN F. *et al.*, "Towards the Monte Carlo simulation of resonant tunnelling diodes using time-dependent wavepackets and Bohm trajectories", *Semiconductor Science and Technology*, vol. 14, no. 6, pp. 532–542, 1999.

[PHI 91] PHILIBERT J., *Atom Movements. Diffusion and Mass Transport in Solids*, Editions de Physique, 1991.

[PIC 80] PICINBONO B., *Introduction à l'étude des signaux et phénomènes aléatoires*, 2nd ed., Dunod, 1980.

[PIC 94] PICKET-MAY M., TAFLOVE A., BARON J., "FDTD-modeling of digital signal propagation in 3-D circuits with active and passive loads", *IEEE Transaction on Microwave Theory Technique*, vol. 42, no. 8, pp. 1514–1523, August 1994.

[QUA 07] QUARTERONI A., SACCO R., SALERI F., *Numerical Methods*, Springer-Verlag Press, Milan, 2007.

[ROH 71] ROHRER R., NAGEL L., MEYER R. *et al.*, "Computationally efficient electronic-circuit noise calculations", *IEEE Journal of Solid-State Circuits*, vol. SC-6, no. 4, p. 204, 1971.

[ROS 98a] ROSSI F., "Topical review on: coherent phenomena in semiconductors", *Semiconductor Science and Technology*, vol. 13, p. 147, 1998.

[ROS 98b] ROSSI F., DI CARLO A., LUGLI P., "Microscopic theory of quantum-transport phenomena in mesoscopic systems: a Monte Carlo approach", *Physical Review Letters*, vol. 80, no.15, pp. 3348–3351, 1998.

[SAD 83] SADANA D.K., SHATAS S.C., GAT A., "Microscopy of semiconducting materials 1983", *Proceedings of the Institute of Physics Conference Series 67*, p. 51, 1983.

[SHI 01] SHIFREN L., FERRY D.K., "Particle Monte Carlo simulation of Wigner function tunneling", *Physics Letters A*, vol. 285, pp. 217–221, 2001.

[SHO 66] SHOCKLEY W., COPELAND J.A., JAMES R.P., "The impedance field method of noise calculation in active semiconductor devices", in LÖWDIN P.O. (ed.), *Quantum Theory of Atoms, Molecules and Solid State*, Academic Press, New York, pp. 537–563, 1966.

[SIL] SILVACO INC., TCAD process, device and EDA: electronic design automation (EDA) software, Santa Clara, California, USA, available at: https://www.silvaco.com.

[SVI 02] SVIZHENKO A., ANANTRAM M.P., GOVINDAN T.R. *et al.*, "Two-dimensional quantum mechanical modeling of nanotransistors", *Journal of Applied Physics*, vol. 91, no. 4, pp. 2343–2354, 2002.

[TAF 05] TAFLOVE A., HAGNESS S., *Computational Electrodynamics: The Finite-Difference Time-Domain Method*, 3rd ed., Artech House, Inc., 2005.

[THO 96] THOBEL J.-L., SLEIMAN A., BOUREL P. *et al.*, "Monte Carlo study of electron transport in III-V heterostructures with doped quantum wells", *Journal of Applied Physics*, vol. 80, no. 2, pp. 928–935, 1996.

[THO 09] THOMPSON P.E. *et al.*, "P and B doped Si resonant interband tunnel diodes with as-grown negative differential resistance", *Electronics Letters*, vol. 45, no. 14, 2009.

[THO 74] THORNBER K.K., MCGILL T.C., NICOLET M.A., "Structure of the Langevin and impedance-field methods of calculating noise in devices", *Solid-State Electronics*, vol. 17, p. 107, 1974.

[TSU 73] TSU R., ESAKI L., "Tunneling in a finite superlattice", *Applied Physics Letters*, vol. 22, pp. 562–564, 1973.

[VAN 70] VAN DER ZIEL A., *Noise. Sources, Characterisation, Measurement*, Prentice Hall, Englewood Cliffs, NJ, 1970.

[VAN 75] VAN VLIET K.M., FRIEDMANN A., ZIJLSTRA R.J.J. *et al.*, "Noise in single injection diodes. I. A survey of methods", *Journal of Applied Physics*, vol. 46, p. 1804, 1975.

[VAN 86] VAN DER ZIEL A., *Noise in Solid State Devices and Circuits*, John Wiley & Sons, 1986

[VER 93] VERGHESE N.K., ALLSTOT D.J., "Rapid simulation of substrate coupling effects in mixed-mode ICs", *Proceedings of the IEEE 1993 Custom Integrated Circuits Conference*, Carnegie Mellon University, Pittsburgh, PA; and S. Masui, Nippon Steel Corp., Sagamihara, Japan, pp. 18.3.1–18.3.4, 1993.

[VIL 02] VILLANI C., "Limites hydrodynamiques de l'équation de Boltzmann", *Séminaires Bourbaki 2000/2001*, vol. 43, pp. 365–405, 2002.

[WAN 07] WANG C.C., KUO C.W., "An efficient scheme for processing arbitrary lumped multiport devices in the finite-difference time-domain method", *IEEE Transaction on Microwave Theory Techniques*, pp. 958–965, 2007.

[WAR 66] WARNER R.M., *Circuits intégrés*, Dunod, Paris, 1966.

[WAT 64] WATKINS G., CORBET J.W., "Defects in irradiated silicon: electron paramagnetic resonance and electron-nuclear double resonance of the Si-E center", *Physical Review Series II*, vol. 134, no. 5a, pp. 342–351, 1964.

[WEN 06] WENBIN X., LAIQING M., WENHUA Y. *et al.*, "A novel expression of lumped elements in FDTD simulations", *International Journal Infrared Millimeters Waves*, vol. 27, no. 7, pp. 985–993, 2006.

[YEE 66] YEE K.S., "Numerical solution of initial boundary value problems involving Maxwell's equations in isotropic media", *IEEE Transaction on Antennas Propagation*, pp. 302–307, 1966.

[YEE 01] YEE X., DREWNAIK J.L., "Incorporating two-port networks with S-parameters into FDTD", *IEEE Microwave and Wireless Components Letters*, vol. 11, no. 2, pp. 77–79, 2001.

Index

A, B, C

D, E, F

G, H, I

L, M, N

P, Q, S

T, V, W

FSC
www.fsc.org
MIX
Papier | Fördert gute Waldnutzung
FSC® C083411

Kolibri 360 GmbH
Ferdinand-Jühlke-Straße 7
99095 Erfurt, Deutschland
produktsicherheit@kolibri360.de